Universitext

*Editorial Board
(North America):*

J.H. Ewing
F.W. Gehring
P.R. Halmos

Universitext

Editors (North America): J.H. Ewing, F.W. Gehring, and P.R. Halmos

Aksoy/Khamsi: Nonstandard Methods in Fixed Point Theory
Aupetit: A Primer on Spectral Theory
Berger: Geometry I, II (two volumes)
Bliedtner/Hansen: Potential Theory
Bloss/Bleeker: Topology and Analysis
Chandrasekharan: Classical Fourier Transforms
Charlap: Bierbach Groups and Flat Manifolds
Chern: Complex Manifolds Without Potential Theory
Cohn: A Classical Invitation to Algebraic Numbers and Class Fields
Curtis: Abstract Linear Algebra
Curtis: Matrix Groups
van Dalen: Logic and Structure
Devlin: Fundamentals of Contemporary Set Theory
Edwards: A Formal Background to Mathematics I a/b
Edwards: A Formal Background to Mathematics II a/b
Foulds: Graph Theory Applications
Emery: Stochastic Calculus
Fukhs/Rokhlin: Beginner's Course in Topology
Frauenthal: Mathematical Modeling in Epidemiology
Gallot/Hulin/Lafontaine: Riemannian Geometry
Gardiner: A First Course in Group Theory
Gårding/Tambour: Algebra for Computer Science
Godbillon: Dynamical Systems on Surfaces
Goldblatt: Orthogonality and Spacetime Geometry
Humi/Miller: Second Course in Order Ordinary Differential Equations
Hurwitz/Kritkos: Lectures on Number Theory
Jones/Morris/Pearson: Abstract Algebra and Famous Impossibilities
Iverson: Cohomology of Sheaves
Kelly/Matthews: The Non-Euclidean Hyperbolic Plain
Kempf: Complex Abelian Varieties and Theta Functions
Kostrikin: Introduction to Algebra
Krasnoselskii/Pekrovskii: Systems with Hypsteresis
Luecking/Rubel: Complex Analysis: A Functional Analysis Approach
Marcus: Number Fields
McCarthy: Introduction to Arithmetical Functions
Meyer: Essential Mathematics for Applied Fields
Mines/Richman/Ruitenburg: A Course in Constructive Algebra
Moise: Introductory Problem Courses in Analysis and Topology
Montesinos: Classical Tessellations and Three Manifolds
Nikulin/Shafarevich: Geometries and Group
Øskendal: Stochastic Differential Equations
Rees: Notes on Geometry
Reisel: Elementary Theory of Metric Spaces
Rey: Introduction to Robust and Quasi-Robust Statistical Methods
Rickart: Natural Function Algebras
Rotman: Galois Theory
Rybakowski: The Homotopy Index and Partial Differential Equations

continued after index

Arthur Jones   Sidney A. Morris
Kenneth R. Pearson

Abstract Algebra and Famous Impossibilities

With 27 Illustrations

Springer-Verlag
New York Berlin Heidelberg London Paris
Tokyo Hong Kong Barcelona Budapest

Arthur Jones
Kenneth R. Pearson
Department of Mathematics
La Trobe University
Bundoora 3083
Australia

Sidney A. Morris
Faculty of Informatics
University of Wollongong
Wollongong 2500
Australia

Editorial Board
(North America):

J.H. Ewing
Department of Mathematics
Indiana University
Bloomington, IN 47405, USA

F.W. Gehring
Department of Mathematics
University of Michigan
Ann Arbor, MI 48109, USA

P.R. Halmos
Department of Mathematics
Santa Clara University
Santa Clara, CA 95053, USA

Library of Congress Cataloging-in-Publication Data
Jones, Arthur, 1934–
    Abstract algebra and famous impossibilities / Arthur Jones, Sidney
A. Morris, Kenneth R. Pearson
      p. cm. — (Universitext)
    Includes bibliographical references and index.
    ISBN 0-387-97661-2
    1. Algebra, Abstract   2. Geometry—Problems, Famous.   I. Morris,
Sidney A., 1947– .   II. Pearson, Kenneth R.   III. Title.
QA162.J65   1992
512'.02—dc20   91-24830

Printed on acid-free paper.

© 1991 Springer-Verlag New York, Inc.
All rights reserved. This work may not be translated or copied in whole or in part without the written permission of the publisher (Springer-Verlag New York, Inc., 175 Fifth Avenue, New York, NY 10010, USA), except for brief excerpts in connection with reviews or scholarly analysis. Use in connection with any form of information storage and retrieval, electronic adaptation, computer software, or by similar or dissimilar methodology now known or hereafter developed is forbidden.
The use of general descriptive names, trade names, trademarks, etc., in this publication, even if the former are not especially identified, is not to be taken as a sign that such names, as understood by the Trade Marks and Merchandise Marks Act, may accordingly be used freely by anyone.

Camera-ready copy provided by the authors.
Printed and bound by R.R. Donnelley and Sons, Harrisonburg, VA.
Printed in the United States of America.

9 8 7 6 5 4 3 2 1

ISBN 0-387-97661-2 Springer-Verlag New York Berlin Heidelberg
ISBN 3-540-97661-2 Springer-Verlag Berlin Heidelberg New York

Preface

The famous problems of squaring the circle, doubling the cube and trisecting an angle captured the imagination of both professional and amateur mathematicians for over two thousand years. Despite the enormous effort and ingenious attempts by these men and women, the problems would not yield to purely geometrical methods. It was only the development of abstract algebra in the nineteenth century which enabled mathematicians to arrive at the surprising conclusion that these constructions are not possible.

In this book we develop enough abstract algebra to prove that these constructions are impossible. Our approach introduces all the relevant concepts about fields in a way which is more concrete than usual and which avoids the use of quotient structures (and even of the Euclidean algorithm for finding the greatest common divisor of two polynomials). Having the geometrical questions as a specific goal provides motivation for the introduction of the algebraic concepts and we have found that students respond very favourably.

We have used this text to teach second-year students at La Trobe University over a period of many years, each time refining the material in the light of student performance.

The text is pitched at a level suitable for students who have already taken a course in linear algebra, including the ideas of a vector space over a field, linear independence, basis and dimension. The treatment, in such a course, of fields and vector spaces as algebraic objects should provide an adequate background for the study of this book. Hence the book is suitable for Junior/Senior courses in North America and second-year courses in Australia.

Chapters 1 to 6, which develop the link between geometry and algebra, are the core of this book. These chapters contain a complete solution to the three famous problems, except for proving that π is a transcendental number (which is needed to complete the proof of the impossibility of squaring the circle). In Chapter 7 we give a self-contained proof that π is transcendental. Chapter 8 contains material about fields which is closely related to the topics in Chapters 2–4, although it is not required in the proof of the impossibility of the three constructions. The short concluding Chapter 9 describes some other areas of mathematics in which algebraic machinery can be used to prove impossibilities.

We expect that any course based on this book will include all of Chapters 1–6 and (ideally) at least passing reference to Chapter 9. We have often taught such a course which we cover in a term (about twenty hours). We find it essential for the course to be paced in a way that allows time for students to do a substantial number of problems for themselves. Different semester length (or longer) courses including topics from Chapters 7 and 8 are possible. The three natural parts of these are

(1) Sections 7.1 and 7.2 (transcendence of e),
(2) Sections 7.3 to 7.6 (transcendence of π),
(3) Chapter 8.

These are independent except, of course, that (2) depends on (1). Possible extensions to the basic course are to include one, two or all of these. While most treatments of the transcendence of π require familiarity with the theory of functions of a complex variable and complex integrals, ours in Chapter 7 is accessible to students who have completed the usual introductory real calculus course (first-year in Australia and Freshman/Sophomore in North America). However instructors should note that the arguments in Sections 7.3 to 7.6 are more difficult and demanding than those in the rest of the book.

Problems are given at the end of each section (rather than collected at the end of the chapter). Some of these are computational and others require students to give simple proofs.

Each chapter contains additional reading suitable for students and instructors. We hope that the text itself will encourage students to do further reading on some of the topics covered.

As in many books, exercises marked with an asterisk * are a good bit harder than the others. We believe it is important to identify clearly the end of each proof and we use the symbol ∎ for this purpose.

We have found that students often lack the mathematical maturity required to write or understand simple proofs. It helps if students write down where the proof is heading, what they have to prove and how they might be able to prove it. Because this is not part of the formal proof, we indicate this exploration by separating it from the proof proper by using a box which looks like

(Include here what must be proved etc.)

Experience has shown that it helps students to use this material if important theorems are given specific names which suggest their content. We have enclosed these names in square brackets before the statement of the theorem. We encourage students to use these names when justifying their solutions to exercises. They often find it convenient to abbreviate the names to just the relevant initials. (For example, the name "Small Degree Irreducibility Theorem" can be abbreviated to S.D.I.T.)

We are especially grateful to our colleague Gary Davis, who pointed the way towards a more concrete treatment of field extensions (using residue rings rather than quotient rings) and thus made the course accessible to a wider class of students. We are grateful to Ernie Bowen, Jeff Brooks, Grant Cairns, Mike Canfell, Brian Davey, Alistair Gray, Paul Halmos, Peter Hodge, Alwyn Horadam, Deborah King, Margaret McIntyre, Bernhard H. Neumann, Kristen Pearson, Suzanne Pearson, Alf van der Poorten, Brailey Sims, Ed Smith and Peter Stacey, who have given us helpful feedback, made suggestions and assisted with the proof reading.

We thank Dorothy Berridge, Ernie Bowen, Helen Cook, Margaret McDonald and Judy Storey for skilful TEXing of the text and diagrams, and Norman Gaywood for assisting with the index.

A.J., S.A.M., K.R.P.
April 1991

Contents

Preface   v

Introduction   1
- 0.1 Three Famous Problems 1
- 0.2 Straightedge and Compass Constructions 3
- 0.3 Impossibility of the Constructions 3

Chapter 1   Algebraic Preliminaries   7
- 1.1 Fields, Rings and Vector Spaces 8
- 1.2 Polynomials 13
- 1.3 The Division Algorithm 17
- 1.4 The Rational Roots Test 20
-       Appendix to Chapter 1 24

Chapter 2   Algebraic Numbers and Their Polynomials   27
- 2.1 Algebraic Numbers 28
- 2.2 Monic Polynomials 31
- 2.3 Monic Polynomials of Least Degree 32

Chapter 3   Extending Fields   39
- 3.1 An Illustration: $\mathbb{Q}(\sqrt{2})$ 40
- 3.2 Construction of $\mathbb{F}(\alpha)$ 44
- 3.3 Iterating the Construction 50
- 3.4 Towers of Fields 52

Chapter 4   Irreducible Polynomials   61
- 4.1 Irreducible Polynomials 62
- 4.2 Reducible Polynomials and Zeros 64
- 4.3 Irreducibility and $\mathrm{irr}(\alpha, \mathbb{F})$ 68
- 4.4 Finite-dimensional Extensions 71

Chapter 5	**Straightedge and Compass Constructions**	**75**
5.1	Standard Straightedge and Compass Constructions	76
5.2	Products, Quotients, Square Roots	85
5.3	Rules for Straightedge and Compass Constructions	89
5.4	Constructible Numbers and Fields	93
Chapter 6	**Proofs of the Impossibilities**	**99**
6.1	Non-Constructible Numbers	100
6.2	The Three Constructions are Impossible	103
6.3	Proving the "All Constructibles Come From Square Roots" Theorem	108
Chapter 7	**Transcendence of e and π**	**115**
7.1	Preliminaries	116
7.2	e is Transcendental	124
7.3	Preliminaries on Symmetric Polynomials	134
7.4	π is Transcendental – Part 1	146
7.5	Preliminaries on Complex-valued Integrals	149
7.6	π is Transcendental – Part 2	153
Chapter 8	**An Algebraic Postscript**	**163**
8.1	The Ring $\mathbb{F}[X]_{p(X)}$	164
8.2	Division and Reciprocals in $\mathbb{F}[X]_{p(X)}$	165
8.3	Reciprocals in $\mathbb{F}(\alpha)$	171
Chapter 9	**Other Impossibilities and Abstract Algebra**	**177**
9.1	Construction of Regular Polygons	178
9.2	Solution of Quintic Equations	179
9.3	Integration in Closed Form	181
Index		**183**

Introduction

0.1 Three Famous Problems

In this book we discuss three of the oldest problems in mathematics. Each of them is over 2,000 years old. The three problems are known as:

[I] doubling the cube (or duplicating the cube, or the Delian problem);

[II] trisecting an arbitrary angle;

[III] squaring the circle (or quadrature of the circle).

Problem I is to construct a cube having twice the volume of a given cube. Problem II is to describe how every angle can be trisected. Problem III is that of constructing a square whose area is equal to that of a given circle. In all cases, the constructions are to be carried out using only a ruler and compass.

Reference to Problem I occurs in the following ancient document supposedly written by Eratosthenes to King Ptolemy III about the year 240 B.C.:

To King Ptolemy, Eratosthenes sends greetings. It is said that one of the ancient tragic poets represented Minos as preparing a tomb for Glaucus and as declaring, when he learnt it was a hundred feet each way: "Small indeed is the tomb thou hast chosen for a royal burial. Let it be double [in volume]. And thou shalt not miss that fair form if thou quickly doublest each side of the tomb." But he was wrong. For when the sides are doubled, the surface [area] becomes four times as great, and the volume eight times. It became a subject of inquiry among geometers in what manner one might double the given volume without changing the shape. And this problem was called the duplication of the cube, for given a cube they sought to double it ...

The origins of Problem II are obscure. The Greeks were concerned with the problem of constructing regular polygons, and it is likely that the trisection problem arose in this context. This is so because the construction of a regular polygon with nine sides necessitates the trisection of an angle.

The history of Problem III is linked to that of calculating the area of a circle. Information about this is contained in the Rhind Papyrus, perhaps the best known ancient mathematical manuscript, which was brought by A.H. Rhind to the British Museum in the nineteenth century. The manuscript was copied by the scribe Ahmes about 1650 B.C. from an even older work. It states that the area of a circle is equal to that of a square whose side is the diameter diminished by one ninth; that is, $A = \left(\frac{8}{9}\right)^2 d^2$. Comparing this with the formula $A = \pi r^2 = \pi \frac{d^2}{4}$ gives

$$\pi = 4 \cdot \left(\frac{8}{9}\right)^2 = \frac{256}{81} = 3.1604\ldots.$$

The Papyrus contains no explanation of how this formula was obtained. Fifteen hundred years later Archimedes showed that

$$3\frac{10}{71} < \pi < 3\frac{10}{70} = 3\frac{1}{7}.$$

(Note that $3\frac{10}{71} = 3.14084\ldots$, $3\frac{1}{7} = 3.14285\ldots$, $\pi = 3.14159\ldots$.)

Throughout the ages these problems were tackled by most of the best mathematicians. For some reason, amateur mathematicians were also fascinated by them. In the time of the Greeks a special word was used to describe people who tried to solve Problem III – τετραγωνιζειν (tetragonidzein) which means *to occupy oneself with the quadrature*.

In 1775 the Paris Academy found it necessary to protect its officials from wasting their time and energy examining purported solutions of these problems by amateur mathematicians. It passed a resolution (Histoire de l'Académie royale, année 1775, p. 61) that no more solutions were to be examined of the problems of doubling the cube, trisecting an arbitrary angle, and squaring the circle and that the same resolution should apply to machines for exhibiting perpetual motion. (See [EH, p. 3].)

The problems were finally solved in the nineteenth century. In 1837, Wantzel settled Problems I and II. In 1882, Lindemann disposed of Problem III.

0.2 Straightedge and Compass Constructions

Construction problems are, and have always been, a favourite topic in geometry. Using only a ruler and compass, a great variety of constructions is possible. Some of these constructions are described in detail in Section 5.1:

a line segment can be bisected; any angle can be bisected; a line can be drawn from a given point perpendicular to a given line; etc.

In all of these problems *the ruler is used merely as a straight edge, an instrument for drawing a straight line but not for measuring or marking off distances.* Such a ruler will be referred to as a *straightedge*.

Problem I is that of constructing with compass and straightedge a cube having twice the volume of a given cube. If the side of the given cube has length 1 unit, then the volume of the given cube is $1^3 = 1$. So the volume of the larger cube should be 2, and its sides should thus have length $\sqrt[3]{2}$. *Hence the problem is reduced to that of constructing, from a segment of length 1, a segment of length $\sqrt[3]{2}$.*

Problem II is to produce a construction for trisecting any given angle. While it is easy to give examples of particular angles which can be trisected, the problem is to give a construction which will work for every angle.

Problem III is that of constructing with compass and straightedge a square of area equal to that of a given circle. If the radius of the circle is taken as one unit, the area of the circle is π, and therefore the area of the constructed square should be π; that is, the side of the square should be $\sqrt{\pi}$. *So the problem is reduced to that of constructing, from a segment of length 1, a segment of length $\sqrt{\pi}$.*

0.3 Impossibility of the Constructions

Why did it take so many centuries for these problems to be solved? The reasons are (i) the required constructions are **impossible**, and (ii) a full understanding of these problems comes not from geometry but from abstract algebra (a subject not born until the nineteenth century). The purpose of this book is to introduce this algebra and show how it is used to prove the impossibility of these constructions.

A real number γ is said to be *constructible* if, starting from a line segment of length 1, we can construct a line segment of length $|\gamma|$ in a finite number of steps using straightedge and compass.

We shall prove, in Chapters 5 and 6, that a real number is constructible if and only if it can be obtained from the number 1 by successive applications of the operations of addition, subtraction, multiplication, division, and taking square roots. Thus, for example, the number
$$2 + \sqrt{3 + 10\sqrt{2}}$$
is constructible.

Now $\sqrt[3]{2}$ does not appear to have this form. Appearances can be deceiving, however. How can we be sure? The answer turns out to be that if $\sqrt[3]{2}$ did have this form, then a certain vector space would have the wrong dimension! This settles Problem I.

As for Problem II, note that it is sufficient to give just one example of an angle which cannot be trisected. One such example is the angle of 60°. It can be shown that this angle can be trisected only if $\cos 20°$ is a constructible number. But, as we shall see in Chapter 6, the number $\cos 20°$ is a solution of the cubic equation
$$8x^3 - 6x - 1 = 0$$
which does not factorize over the rational numbers. Hence it seems likely that *cube roots*, rather than square roots, will be involved in its solution, so we would not expect $\cos 20°$ to be constructible. Once again this can be made into a rigorous proof by considering the possible dimensions of a certain vector space.

As we shall show in Chapter 6, the solution of Problem III also hinges on the dimension of a vector space. Indeed, the impossibility of squaring the circle follows from the fact that a certain vector space (a different one from those mentioned above in connection with Problems I and II) is not finite-dimensional. This in turn is because the number π is "transcendental", which we shall prove in Chapter 7.

Additional Reading for the Introduction

More information about the background to the three problems can be found in the various books on the history of mathematics listed below. References dealing specifically with Greek mathematics include [JG], [TH] and [FLa]. A detailed history of Problem III is given in [EH].

Introduction

The original solutions to Problems I and II by Wantzel are in [MW] and the original solution to Problem III by Lindemann is in [FL].

[EB1] E.T. Bell, *Men of Mathematics,* Simon and Shuster, New York, 1937.

[EB2] E.T. Bell, *The Development of Mathematics,* McGraw-Hill, New York, 1945.

[CB] C.B. Boyer, *A History of Mathematics,* Wiley, New York, 1968.

[RC] R. Courant and H. Robbins, *What is Mathematics?,* Oxford University Press, New York, 1941.

[AD] A. De Morgan, *A Budget of Paradoxes,* Dover, New York, 1954.

[UD] U. Dudley, *A Budget of Trisections,* Springer-Verlag, New York, 1987.

[JG] J. Gow, *A Short History of Greek Mathematics,* Chelsea, New York, 1968.

[CH] C.R. Hadlock, *Field Theory and its Classical Problems,* Carus Mathematical Monographs, No. 19, Mathematical Association of America, 1978.

[TH] T.L. Heath, *A History of Greek Mathematics Vol.I & II,* Clarendon Press, Oxford, 1921.

[EH] E.W. Hobson, *Squaring the Circle,* Cambridge University Press, 1913; reprinted in *Squaring the Circle, and Other Monographs,* Chelsea, 1953.

[HH] H.P. Hudson, *Ruler and Compass,* Longmans Green, 1916; reprinted in *Squaring the Circle, and Other Monographs,* Chelsea, 1953.

[FK] F. Klein, *Famous Problems, and other monographs,* Chelsea, New York, 1962.

[MK] M. Kline, *Mathematical Thought from Ancient to Modern Times,* Oxford University Press, 1972.

[FLa] F. Lasserre, *The Birth of Mathematics in the Age of Plato,* Hutchinson, London, 1964.

[FL] F.L. Lindemann, "Über die Zahl π", *Mathematische Annalen,* 20 (1882), 213–225.

[DS] D.J. Struik, *A Concise History of Mathematics,* Dover, New York, 1967.

[VS] V. Sanford, *A Short History of Mathematics,* Harrap, London, 1958.

[HT] H.W. Turnbull, *The Great Mathematicians,* Methuen, London 1933.

[MW] M.L. Wantzel, "Recherches sur les moyens de reconnaître si un Problème de Géométrie peut se résoudre avec la règle et le compas", *Journal de Mathématiques Pures et Appliquées,* 2 (1837), 366-372.

CHAPTER 1

Algebraic Preliminaries

This chapter presents the background algebra on which the rest of this book depends. Much of this material should be familiar to you.

The Rational Roots Test, however, will probably be new to you. It provides a nice illustration of how polynomials can be used to study certain properties of real numbers; for example, we use it to show that the numbers $\sqrt[5]{2}$, $\sqrt{2}+\sqrt{3}$, and $\sin 20°$ are irrational. The application of polynomials to the study of numbers will be an important theme throughout the book.

1.1 Fields, Rings and Vector Spaces

In this section we summarize the main ideas and terminology of fields, rings, and vector spaces which we shall use throughout this book. If you are not already familiar with some of this material, we refer you to the Appendix to this chapter for more details.

Fields

A familiar example of a *field* is the set $\mathbb{Q}$ of all rational numbers. For this set, the usual operations of arithmetic

addition, subtraction, multiplication, and division (except by 0)

can be performed without restriction. The same is true of the set $\mathbb{R}$ of all real numbers, which is another example of a field. Likewise the set $\mathbb{C}$ of all complex numbers, with the above operations, is a field.

A formal definition of a field is given in the Appendix to this chapter. Strictly speaking, a field consists of a set $\mathbb{F}$ together with the operations to be performed on $\mathbb{F}$. When there is no ambiguity as to which operations are intended, we simply refer to the set $\mathbb{F}$ as the field.

If $\mathbb{F}$ is a field and $\mathbb{E} \subseteq \mathbb{F}$ it may happen that $\mathbb{E}$ also becomes a field when we apply the operations on $\mathbb{F}$ to its elements. If this happens we say that $\mathbb{E}$ is a *subfield* of $\mathbb{F}$ and, reciprocally, that $\mathbb{F}$ is an *extension field* of $\mathbb{E}$. Thus $\mathbb{Q}$ is a subfield of both $\mathbb{C}$ and $\mathbb{R}$ while $\mathbb{R}$ and $\mathbb{C}$ are both extension fields of $\mathbb{Q}$.

An important relationship between $\mathbb{Q}$ and $\mathbb{C}$ is expressed in the following proposition.

1.1.1 Proposition. *$\mathbb{Q}$ is the smallest subfield of $\mathbb{C}$.*

Proof. We are to show that if $\mathbb{F}$ is any subfield of $\mathbb{C}$ then $\mathbb{Q} \subseteq \mathbb{F}$. So we let $\mathbb{F}$ be any subfield of $\mathbb{C}$.

To prove that $\mathbb{Q} \subseteq \mathbb{F}$ we shall show that

$$\text{if} \quad x \in \mathbb{Q} \quad \text{then} \quad x \in \mathbb{F}.$$

Let $x \in \mathbb{Q}$; that is,

$$x = \frac{r}{s}, \quad \text{where } r, s \text{ are integers and } s \neq 0.$$

Algebraic Preliminaries

> Our aim is to prove that $x \in \mathsf{F}$.
> To do this we use the field properties of F.

As F is a field, the number 1 must be in F.
Therefore each positive integer is in F, as F is closed under addition.
So each negative integer is in F, as F is closed under subtraction.
Also $0 \in \mathsf{F}$, as F is a field.
Hence the integers r and s are in F.
Since F is closed under division (and $s \neq 0$), the quotient $r/s \in \mathsf{F}$.
So $x \in \mathsf{F}$, as required. ∎

There are, in fact, lots of interesting fields which lie between $\mathbb{Q}$ and $\mathbb{C}$ and we shall devote much time to studying them.

Of course not all fields are subfields of $\mathbb{C}$. To see this, consider the following. For each positive integer n put

$$\mathbb{Z}_n = \{0, 1, 2, \ldots, n-1\}.$$

We define addition of elements a and b in $\mathbb{Z}_n$ by

$$a \oplus_n b = a + b \pmod{n};$$

that is, add a and b in the usual way and then subtract multiples of n until the answer lies in the set $\mathbb{Z}_n$. We define multiplication similarly:

$$a \otimes_n b = a \times b \pmod{n}.$$

For example,

$$3 \oplus_6 4 = 1 \quad \text{(subtract 6 from 7)},$$
$$2 \oplus_6 1 = 3,$$
$$3 \otimes_6 4 = 0 \quad \text{(subtract 12 from 12)},$$
$$9 \otimes_{10} 7 = 3.$$

It can be shown that if n is a prime number then $\mathbb{Z}_n$ is a field. It is obvious that in this case $\mathbb{Z}_n$ is not a subfield of $\mathbb{C}$. For example, if $n = 2$, $1 \oplus_2 1 = 0$ in $\mathbb{Z}_2$, but $1 + 1 = 2$ in $\mathbb{C}$; so the operation of addition in the field $\mathbb{Z}_2$ is different from that in the field $\mathbb{C}$. Unlike our previous examples $\mathbb{Q}$, $\mathbb{R}$ and $\mathbb{C}$, the field $\mathbb{Z}_n$ is a *finite field*; that is, it has only a finite number of elements.

Rings

The concept of a ring is less restrictive than that of a field in that we no longer require the possibility of division but only of

addition, subtraction, and multiplication.

Thus the set $\mathbb{Z}$ of all the integers is an example of a ring which is not a field, whereas the set $\mathbb{N}$ of all natural numbers is not a ring.

We even allow the concept of multiplication itself to be liberalized. We do not require it to be commutative; that is, ab may be unequal to ba. We also permit zero-divisors, so that it is possible to have

$ab = 0$ *but neither a nor b is zero.*

Observe that if M is the set of all 2×2 matrices with entries from $\mathbb{R}$, then M is a ring with the ring operations being matrix addition and matrix multiplication. However, M is not a field since if

$$A = \begin{pmatrix} 1 & 2 \\ 3 & 4 \end{pmatrix} \quad \text{and} \quad B = \begin{pmatrix} 5 & 6 \\ 7 & 8 \end{pmatrix}$$

$$\text{then} \quad AB = \begin{pmatrix} 19 & 22 \\ 43 & 50 \end{pmatrix} \neq BA = \begin{pmatrix} 23 & 34 \\ 31 & 46 \end{pmatrix}.$$

Note also that this ring M has zero-divisors; for example, if

$$C = \begin{pmatrix} 1 & 0 \\ 0 & 0 \end{pmatrix} \quad \text{and} \quad D = \begin{pmatrix} 0 & 0 \\ 4 & 5 \end{pmatrix} \quad \text{then} \quad CD = \begin{pmatrix} 0 & 0 \\ 0 & 0 \end{pmatrix}.$$

A formal definition of a ring is given in the Appendix to this chapter.

Vector Spaces

Recall from linear algebra that a *vector space* consists of a set V together with a field $\mathbb{F}$. The elements of V are called *vectors* and the elements of $\mathbb{F}$ are called *scalars*. Any two vectors can be added to give a vector and a vector can be multiplied by a scalar to give a vector. A formal definition of a vector space is given in the Appendix to this chapter.

When there is no danger of ambiguity it is customary to omit reference to both the scalars and the operations and to regard V itself as the vector space. In our study, however, the field of scalars will be constantly changing and to omit reference to it would almost certainly lead to ambiguity. Hence we shall include the field $\mathbb{F}$ in our description and refer to *"the vector space V over $\mathbb{F}$"*.

Algebraic Preliminaries

Two other notions which you need to recall are those of spanning and linear independence.

If $v_1, v_2, \ldots, v_n$ are vectors in the vector space V over the field $\mathbb{F}$, then we say that the set $\{v_1, v_2, \ldots, v_n\}$ of these vectors *spans* V *over* $\mathbb{F}$ providing that every vector $v \in V$ can be written as a linear combination of $v_1, v_2, \ldots, v_n$; that is,

$$v = \lambda_1 v_1 + \lambda_2 v_2 + \ldots + \lambda_n v_n$$

for some $\lambda_1, \lambda_2, \ldots, \lambda_n$ in $\mathbb{F}$.

A vector space is said to be *finite-dimensional* if there is a finite set of vectors $\{v_1, v_2, \ldots, v_n\}$ which spans it.

The set $\{v_1, v_2, \ldots, v_n\}$ of vectors is said to be *linearly independent over* $\mathbb{F}$ if the zero vector, 0, can be written only as a trivial linear combination of $v_1, v_2, \ldots, v_n$; that is,

$$0 = \lambda_1 v_1 + \lambda_2 v_2 + \ldots + \lambda_n v_n \qquad (\lambda_i \in \mathbb{F})$$

implies that $\lambda_1 = \lambda_2 = \ldots = \lambda_n = 0$.

The set $\{v_1, v_2, \ldots, v_n\}$ is said to be a *basis* for V over $\mathbb{F}$ if it is linearly independent over $\mathbb{F}$ and spans V over $\mathbb{F}$.

While a finite-dimensional vector space can have an infinite number of distinct bases, the number of vectors in each of the bases is the same and is called the *dimension* of the vector space. Recall also that, in a vector space of dimension n, any set with more than n vectors must be linearly dependent. (For example, any set of four vectors in a three-dimensional space is linearly dependent.)

Pairs of Fields

The way in which vector spaces enter our study is as pairs of fields, one of which is a subfield of the other.

The most intuitive example, in this regard, is the vector space $\mathbb{C}$ over $\mathbb{R}$ in which we take $\mathbb{C}$ for the vectors and $\mathbb{R}$ for the scalars. This vector space arises from the pair of fields $\mathbb{C}$ and $\mathbb{R}$, in which $\mathbb{R}$ is a subfield of $\mathbb{C}$. We may think of the elements of $\mathbb{C}$ as points in the plane and then the vector space operations correspond exactly to the usual vector addition and scalar multiplication in the plane.

More generally, if $\mathbb{E}$ and $\mathbb{F}$ are fields with $\mathbb{E}$ a subfield of $\mathbb{F}$, we may take $\mathbb{F}$ for the vectors and $\mathbb{E}$ for the scalars to get the *vector space* $\mathbb{F}$ *over* $\mathbb{E}$. (Note that the smaller field is the scalars and the larger one is the vectors.)

1.1.2 Definition. If the vector space $\mathbb{F}$ over $\mathbb{E}$ is finite-dimensional its dimension is denoted by $[\mathbb{F} : \mathbb{E}]$ and called the *degree of* $\mathbb{F}$ *over* $\mathbb{E}$. ∎

For example, it is readily seen that $[\mathbb{C} : \mathbb{R}] = 2$ and $[\mathbb{R} : \mathbb{R}] = 1$.

──────────────── Exercises 1.1 ────────────────

1. (a) Which of the following are meaningful?
    (i) the vector space $\mathbb{C}$ over $\mathbb{R}$;
    (ii) the vector space $\mathbb{R}$ over $\mathbb{R}$;
    (iii) the vector space $\mathbb{R}$ over $\mathbb{C}$;
    (iv) the vector space $\mathbb{C}$ over $\mathbb{Q}$;
    (v) the vector space $\mathbb{C}$ over $\mathbb{C}$;
    (vi) the vector space $\mathbb{Q}$ over $\mathbb{Q}$;
    (vii) the vector space $\mathbb{Q}$ over $\mathbb{R}$;
    (viii) the vector space $\mathbb{Q}$ over $\mathbb{C}$.
   (b) Which of the above vector spaces has dimension 2? Write down a basis in each such case.

2. How may subfields of $\mathbb{Q}$ are there? (Justify your answer.)

3. (a) Is $\{(1 + i\sqrt{2}), (\sqrt{2} + 2i)\}$ a linearly independent subset of the vector space $\mathbb{C}$ over $\mathbb{R}$?
   (b) Is $\{(1 + i\sqrt{2}), (\sqrt{2} + 2i)\}$ a linearly independent subset of the vector space $\mathbb{C}$ over $\mathbb{Q}$?

4. (a) Let $M_2(\mathbb{R})$ be the set of all 2×2 matrices with entries from $\mathbb{R}$. What are the natural vector space addition and scalar multiplication which make $M_2(\mathbb{R})$ a vector space over $\mathbb{R}$?
   (b) Find a basis for $M_2(\mathbb{R})$.
   (c) Let $M_2(\mathbb{C})$ be the set of all 2×2 matrices with entries from $\mathbb{C}$. Show that it is a vector space over $\mathbb{R}$ and find a basis for it. Is it also a vector space over $\mathbb{C}$?

5. Let $\mathbb{F} = \{a + b\sqrt{2} : a, b \in \mathbb{Q}\}$.
   (a) Prove that $\mathbb{F}$ is a vector space over $\mathbb{Q}$ and write down a basis for it.
   (b)* Prove that $\mathbb{F}$ is a field.
   (c) Find the value of $[\mathbb{F} : \mathbb{Q}]$.

Algebraic Preliminaries

6. A ring R is said to be an *integral domain* if its multiplication is commutative, if there is an element 1 such that $1.x = x.1 = x$ for all $x \in R$, and if there are no zero-divisors.

   (a) Is $\mathbb{Z}_4$ an integral domain?

   (b)* For what $n \in \mathbb{N}$ is $\mathbb{Z}_n$ an integral domain?

7. Let $\mathbb{E}$ be a subfield of a finite field $\mathbb{F}$ such that $[\mathbb{F} : \mathbb{E}] = n$, for some $n \in \mathbb{N}$. If $\mathbb{E}$ has m elements, for some $m \in \mathbb{N}$, how many elements does $\mathbb{F}$ have? (Justify your answer.)

8. (a) If A, B, and C are finite fields with A a subfield of B and B a subfield of C, prove that $[C : A] = [C : B].[B : A]$.

   [Hint. Use the result of Exercise 7.]

   (b) If furthermore $[C : A]$ is a prime number, deduce that $B = C$ or $B = A$.

1.2 Polynomials

The simplest and most naive way of describing a polynomial is to say that it is an "expression" of the form

$$a_0 + a_1 x + a_2 x^2 + \ldots + a_n x^n . \tag{1}$$

The coefficients $a_0, a_1, \ldots, a_n$ are assumed to belong to some ring R. The above expression is then called a *polynomial over the ring R*.

As yet we have said nothing about the symbol x appearing in the above expression. There are, in fact, two distinct ways of interpreting this symbol. One of these ways leads to the concept of a *polynomial form*, the other to that of a *polynomial function*.

Polynomial Forms

To arrive at the concept of a polynomial form, we say as little as possible about the symbol "x". We regard x and its various powers as simply performing the rôle of "markers" to indicate the position of the various coefficients in the expression.

Thus, for example, we know that the two polynomials

$$1 + 2x + 3x^2 \quad \text{and} \quad 2x + 3x^2 + 1$$

are equal because in each expression the coefficients of the same powers of x are equal.

Again, when we add two polynomials, as in the following calculation,
$$(1 + 2x + 3x^2) + (2 + x^2) = 3 + 2x + 4x^2,$$
we may regard the powers of x as simply telling us which coefficients to add together. They play a similar, although more complicated, rôle when we multiply two polynomials.

Although x is regarded as obeying the same algebraic rules as the elements of the ring R, we do not think of it as assuming values from R. For this reason it is called an *indeterminate*.

For emphasis, we shall use upper case letters for indeterminates in this book and write "X" instead of "x" so our polynomial (1) becomes
$$a_0 + a_1 X + a_2 X^2 + \ldots + a_n X^n,$$
and we call it a *polynomial form* (over the ring R, in the indeterminate X).

What really matters in a polynomial form are the coefficients. Accordingly we say that *two polynomial forms are equal if and only if they have the same coefficients.*

1.2.1 Definitions. If in the above polynomial form the last coefficient $a_n \neq 0$, then we say that the polynomial has *degree n*.

If all the coefficients are zero, we say that the polynomial is the *zero polynomial* and leave its degree undefined. ∎

When we do not wish to state the coefficients explicitly we shall use symbols like $f(X), g(X), h(X)$ to denote polynomials in X. The degree of a nonzero polynomial $f(X)$ is denoted by
$$\deg f(X).$$

If in the above polynomial form $a_n \neq 0$, we call a_n the *leading coefficient*.

The collection of all polynomials over the ring R in the indeterminate X will be denoted by
$$R[X]$$
(note the square brackets, as distinct from round ones). Since polynomial forms can be added, subtracted and multiplied (and these operations obey the usual algebraic laws) this set of polynomials is itself a ring.

Algebraic Preliminaries

In particular, if the ring R happens to be a field $\mathbb{F}$, we obtain the polynomial ring $\mathbb{F}[X]$. This ring will not be a field, however, since $X \in \mathbb{F}[X]$ but X has no reciprocal in $\mathbb{F}[X]$ because

$$X \cdot f(X) \neq 1, \quad \text{for all } f(X) \in \mathbb{F}[X].$$

Polynomial Functions

Back in the original expression (1), an alternative way to regard the symbol "x" is as a variable standing for a typical element of the ring R. Thus the expression (1) may be used to assign to each element $x \in R$ another element in R. In this way we get a function $f \colon R \to R$ with values assigned by the formula

$$f(x) = a_0 + a_1 x + a_2 x^2 + \ldots + a_n x^n. \tag{2}$$

Such a function f is called a *polynomial function* on the ring R.

Thus if we regard polynomials as functions, emphasis shifts from the coefficients to the values of the function. In particular, *equality of two polynomial functions* $f \colon R \to R$ and $g \colon R \to R$ means that

$$f(x) = g(x), \quad \text{for all } x \in R,$$

which is just the standard definition of equality for functions.

Clearly each polynomial form $f(X)$ determines a unique polynomial function $f \colon R \to R$ (because we can read off the coefficients from $f(X)$ and use them to generate the values of f via the formula (2)). Is the converse true? *Does each polynomial function* $f \colon R \to R$ *determine a unique polynomial form?* The answer is: *not always!* The following example shows why.

1.2.2 Example. Let $f(X)$ and $g(X)$ be the polynomial forms over the ring $\mathbb{Z}_4$ given by

$$f(X) = 2X \quad \text{and} \quad g(X) = 2X^2.$$

These two polynomial forms are different, yet they determine the same polynomial function.

Proof. These polynomial forms determine functions $f \colon \mathbb{Z}_4 \to \mathbb{Z}_4$ and $g \colon \mathbb{Z}_4 \to \mathbb{Z}_4$, with values given by

$$f(x) = 2x \quad \text{and} \quad g(x) = 2x^2.$$

Hence calculation in $\mathbb{Z}_4$ shows that

$$f(0) = 0 = g(0)$$
$$f(1) = 2 = g(1)$$
$$f(2) = 0 = g(2)$$
$$f(3) = 2 = g(3).$$

So $f(x) = g(x)$, for all $x \in \mathbb{Z}_4$. Thus the functions f and g are equal. ∎

This is something of an embarrassment! We have produced an example of a ring R and two polynomial forms $f(X), g(X) \in R[X]$ such that

$$f(X) \neq g(X) \quad \text{and yet} \quad f = g.$$

Fortunately, however, the only rings which are relevant to our geometrical construction problems are those which are subfields of the complex number field, $\mathbb{C}$. For such rings it can be shown (see Exercises 1.2 #6) that the above phenomenon cannot occur and we get a one-to-one correspondence

$$f(X) \leftrightarrow f$$

between polynomial forms $f(X) \in R[X]$ and polynomial functions $f: R \to R$.

The result of all this is that while there is a fine conceptual distinction between polynomial forms and polynomial functions, we can ignore the distinction in the remainder of this book without running into practical difficulties. *Thus*

$$f(X) \quad \text{and} \quad f$$

will be more or less the same while $f(x)$ will be quite different, being the value of f at x.

──────────────── **Exercises 1.2** ────────────────

1. Give an example of an element of the polynomial ring $\mathbb{R}[X]$ which is not a member of $\mathbb{Q}[X]$.

2. In each of the following cases, give a pair of nonzero polynomials $f(X), g(X) \in \mathbb{Q}[X]$ which satisfies the condition:
   (i) $\deg(f(X) + g(X)) < \deg f(X) + \deg g(X)$;

Algebraic Preliminaries

(ii) $\deg(f(X) + g(X)) = \deg f(X) + \deg g(X)$;

(iii) $\deg(f(X) + g(X))$ is undefined.

3. (a) Let $\mathbb{F}$ be a field. Verify that if $f(X)$ and $g(X)$ are nonzero polynomials in $\mathbb{F}[X]$ then $f(X)g(X) \neq 0$ and

$$\deg f(X)g(X) = \deg f(X) + \deg g(X).$$

(b) Deduce that the ring $\mathbb{F}[X]$ is an integral domain.

(See Exercises 1.1 #6 for the definition of integral domain.)

(c) Is $\mathbb{Z}_6[X]$ an integral domain? (Justify your answer.)

4. Let $f(X)$ denote the polynomial form over the ring $\mathbb{Z}_6$ given by $f(X) = X^3$. Find a polynomial form $g(X)$, with $f(X) \neq g(X)$, such that $f(X)$ and $g(X)$ determine the same polynomial function.

5.* Let R be any finite ring. Prove that there exist unequal polynomial forms $f(X)$ and $g(X)$ over R such that $f(X)$ and $g(X)$ determine the same polynomial function.

6. Let $\mathbb{F}$ be any subfield of the complex number field $\mathbb{C}$ and $f(X)$ and $g(X)$ be polynomial forms over the field $\mathbb{F}$. If the polynomial functions f and g are equal (that is, $f(x) = g(x)$, for all $x \in \mathbb{F}$), prove that the polynomial forms are equal (that is, $f(X) = g(X)$).

[Hint. You may assume the well-known result that for any polynomial function h of degree n, there are at most n complex numbers x such that $h(x) = 0$.]

1.3 The Division Algorithm

When you divide one polynomial by another you get a quotient and a remainder. The procedure for doing this is called the *division algorithm* for polynomials. It leads to a theorem which is of fundamental importance in the study of polynomials. We begin with an example.

1.3.1 Example. We shall divide the polynomial $X^3 + 2X^2 + 3X + 4$ by the polynomial $X + 1$ to produce a quotient and remainder in $\mathbb{Q}[X]$.

$$\begin{array}{r}
X^2 + X + 2 \leftarrow quotient \\
divisor \rightarrow X+1 \overline{\smash{\big)}\, X^3 + 2X^2 + 3X + 4} \leftarrow dividend \\
\underline{X^3 + X^2 } \\
X^2 + 3X + 4 \\
\underline{X^2 + X } \\
2X + 4 \\
\underline{2X + 2 } \\
2 \leftarrow remainder
\end{array}$$

It is easy to check that the algorithm has done its job and given us what we want, namely,

$$dividend = divisor \times quotient + remainder.$$

To understand why the division algorithm works it is helpful to write the results from the various stages of the process as equalities:

$$\left.\begin{array}{r}
X^3 + 2X^2 + 3X + 4 = (X+1)X^2 + (X^2 + 3X + 4) \\
X^2 + 3X + 4 = (X+1)X + (2X + 4) \\
2X + 4 = (X+1)2 + 2
\end{array}\right\} \quad (1)$$

To help understand the significance of the equations (1), let us introduce some names for the various terms occurring in them and hence write each of the equations (1) as

current dividend $= (X+1)$ (*monomial in X*) $+$ *current remainder*.

(A *monomial* in X means a polynomial of the type cX^i.) In the first of the equations (1), the current dividend is the actual dividend. At each step, the aim is to choose the monomial so as to balance out the highest power of X in the current dividend. This will ensure that

$$\deg current\ remainder < \deg current\ dividend.$$

At the next step, the old current remainder becomes the new current dividend. The process stops when eventually (at the third stage in our example) the degree of the current remainder falls below the degree of the divisor ($X + 1$ in this case).

Algebraic Preliminaries

The final step is to put the three equations (1) together to get

$$X^3 + 2X^2 + 3X + 4 = (X+1)(X^2 + X + 2) + 2$$
$$\text{\textit{dividend}} \qquad \text{\textit{divisor}} \quad \text{\textit{quotient}} \qquad \text{\textit{remainder}}$$

thereby verifying that the algorithm has achieved its goal. ∎

The division algorithm can be applied more generally to a dividend $f(X)$ and divisor $g(X)$ to give a quotient $q(X)$ and remainder $r(X)$ as in the following theorem.

1.3.2 Theorem. [Division Theorem] *Let $\mathbb{F}$ be a field. If $f(X)$ and $g(X)$ are in $\mathbb{F}[X]$ and $g(X) \neq 0$, then there are polynomials $q(X)$ and $r(X)$ in $\mathbb{F}[X]$ such that*

$$f(X) = g(X)q(X) + r(X)$$

and either $r(X) = 0$ or $\deg r(X) < \deg g(X)$.

Proof. The division algorithm is used to construct the polynomials $q(X)$ and $r(X)$. The proof that the algorithm succeeds uses the same ideas as those explained in the above example. ∎

--- Exercises 1.3 ---

1. Find the quotient and remainder when $X^3 + 2X + 1$ is divided by $2X + 1$ in $\mathbb{Q}[X]$.

2. (a) Repeat Exercise 1 above with $\mathbb{Z}_7[X]$ replacing $\mathbb{Q}[X]$.
   (b) Write your answer to (a) without using any "−" signs.

3. Use the division algorithm to find the quotient and remainder when $X^3 + iX^2 + 3X + i$ is divided by $X^2 - i$ in $\mathbb{C}[X]$.

4. Use the division algorithm to find the quotient and remainder when $X^2 + \sqrt{2}X - 3$ is divided by $X - \sqrt{3}$ in $\mathbb{R}[X]$.

5.* Write out the proof of Theorem 1.3.2 in detail.
   [Hint. Your proof may use mathematical induction.]

6.* Find an example which shows that Theorem 1.3.2 would be false if we assumed that $\mathbb{F}$ was only a ring rather than a field.

7.* In Theorem 1.3.2 if $\mathbb{F}$, $f(X)$ and $g(X)$ are given, show that $q(X)$ and $r(X)$ are uniquely determined.

1.4 The Rational Roots Test

This simple test will enable you to prove very easily that certain numbers, such as

$$\sqrt{3}, \quad \sqrt[5]{2}, \quad \text{and} \quad \sqrt{2}+\sqrt{3},$$

are irrational. The test illustrates how polynomials can be applied to the study of numbers.

A *zero* of a polynomial $f(X)$ is a number β such that $f(\beta) = 0$. The test works by narrowing down to a short list the possible zeros in $\mathbb{Q}$ of a polynomial in $\mathbb{Q}[X]$.

A preliminary observation is that every polynomial in $\mathbb{Q}[X]$ can be written as a rational multiple of a polynomial in $\mathbb{Z}[X]$. This is achieved by multiplying the given polynomial in $\mathbb{Q}[X]$ by a suitable integer. For example

$$1 + \frac{1}{3}X + \frac{2}{7}X^2 + \frac{1}{2}X^3 = \frac{1}{42}(42 + 14X + 12X^2 + 21X^3).$$

In looking for zeros in $\mathbb{Q}$ of polynomials in $\mathbb{Q}[X]$ we may as well, therefore, look for zeros in $\mathbb{Q}$ of polynomials in $\mathbb{Z}[X]$.

The Rational Roots Test uses some terminology which we now record.

By saying that an integer p is *a factor of* an integer q, we mean that there is an integer m such that $q = mp$.

By saying that a rational number β is expressed in *lowest terms* by $\beta = \frac{r}{s}$ we mean that r and s are in $\mathbb{Z}$ with $s \neq 0$ and r and s have no common factors except 1 and -1.

1.4.1 Theorem. [Rational Roots Test] *Let $f(X) \in \mathbb{Z}[X]$ be a polynomial of degree n so that*

$$f(X) = a_0 + a_1 X + \ldots + a_n X^n$$

for some $a_0, a_1, \ldots, a_n \in \mathbb{Z}$, with $a_n \neq 0$. If β is a rational number, written in its lowest terms as $\beta = \frac{r}{s}$, and β is a zero of $f(X)$ then

(i) *r is a factor of a_0, and*

(ii) *s is a factor of a_n.*

Proof. Let $f(\beta) = 0$ so that

$$a_0 + a_1 \left(\frac{r}{s}\right) + a_2 \left(\frac{r}{s}\right)^2 + \ldots + a_n \left(\frac{r}{s}\right)^n = 0$$

and hence
$$a_0 s^n + a_1 r s^{n-1} + a_2 r^2 s^{n-2} + \ldots + a_n r^n = 0.$$
This gives
$$a_0 s^n = -r\left(a_1 s^{n-1} + a_2 r s^{n-2} + \ldots + a_n r^{n-1}\right)$$
so that
$$r \text{ is a factor of } a_0 s^n.$$
But since r and s have no common factors except 1 and -1, there can be no common prime numbers in the prime factorizations of r and s. This implies that
$$r \text{ is a factor of } a_0.$$
(This follows from the Fundamental Theorem of Arithmetic; see Exercises 1.4 #9.)

Similarly on writing
$$a_n r^n = -s\left(a_0 s^{n-1} + a_1 r s^{n-2} + \ldots + a_{n-1} r^{n-1}\right)$$
we see that
$$s \text{ is a factor of } a_n r^n$$
and therefore
$$s \text{ is a factor of } a_n. \qquad \blacksquare$$

The following example is a typical application of the Rational Roots Test.

1.4.2 Example. The real number $\sqrt[5]{2}$ is not rational.

Proof. Note first that $\sqrt[5]{2}$ is a zero of the polynomial $2 - X^5$ in $\mathbb{Q}[X]$. This polynomial also lies in $\mathbb{Z}[X]$, and hence we can use the Rational Roots Test to see if it has a zero in $\mathbb{Q}$.

Suppose therefore that $\frac{r}{s}$ is a zero of $2 - X^5$, where $r, s \in \mathbb{Z}$ with $s \neq 0$ and $\frac{r}{s}$ is expressed in lowest terms. By the Rational Roots Test,
$$r \text{ is a factor of } 2 \quad \text{and} \quad s \text{ is a factor of } -1.$$
This means that the only possible values of r are $1, -1, 2,$ and -2, while the only possible values of s are 1 and -1. Hence $\frac{r}{s}$ must be $1, -1, 2,$ or -2. This means that ± 1 and ± 2 are the only possible zeros in $\mathbb{Q}$. Substitution shows, however, that none of ± 1 and ± 2 is a zero of $2 - X^5$, and so $2 - X^5$ has no zeros in $\mathbb{Q}$. Hence $\sqrt[5]{2}$, being a zero, is not in $\mathbb{Q}$. $\qquad \blacksquare$

Exercises 1.4

1. Let $p(X)$ be the element of $\mathbb{Q}[X]$ given by
   $$p(X) = 2X^3 + 3X^2 + 2X + 3.$$
   (a) Use the Rational Roots Test to find all possible rational zeros of $p(X)$.
   (b) Is -1 a zero of $p(X)$?
   (c) Is $-\frac{3}{2}$ a zero of $p(X)$?

2. Use the Rational Roots Test to prove that $\sqrt{5}$ is irrational.

3. Find a polynomial in $\mathbb{Q}[X]$ which has $\sqrt{2} + \sqrt{3}$ as a zero. Hence show that $\sqrt{2} + \sqrt{3}$ is irrational.
   [Hint. To obtain the polynomial, first let $\alpha = \sqrt{2} + \sqrt{3}$ and then square both sides.]

4. (a) Use the Rational Roots Test to prove that for each $m \in \mathbb{N}$, $\sqrt{m}$ is rational if and only if m is a perfect square.
   (b) Let m and n be any positive integers. Prove that $\sqrt[n]{m}$ is a rational number if and only if it is an integer.

5. If n is any positive integer, prove that $\sqrt{n} + \sqrt{n+1}$ and $\sqrt{n} - \sqrt{n+1}$ are irrational.

6. Prove that $\sqrt{n+1} - \sqrt{n-1}$ is irrational, for every positive integer n.

7. (a)* Prove that $\sqrt{3} + \sqrt{5} + \sqrt{7}$ is irrational.
   (b)* Prove that $\sqrt{2} + \sqrt{3} + \sqrt{5}$ is irrational.
   [Hint. Both (a) and (b) involve some arithmetic. Do not be afraid to use a calculator or a computer program in an appropriate fashion.]

8. Find a polynomial in $\mathbb{Q}[X]$ with α as a zero, where
   $$\alpha = \sqrt[3]{4} - \sqrt[3]{2}$$
   and then deduce that $\alpha \notin \mathbb{Q}$ by applying the Rational Roots Test.
   [Hint. Calculate α^3 and keep an eye on α at the same time.]

Algebraic Preliminaries

9. The *Fundamental Theorem of Arithmetic* says that every natural number $n > 1$ can be written as a product of prime numbers and, except for the order of the factors, the expression of n in this form is unique. Thus n has a unique expression

$$n = p_1^{a_1} p_2^{a_2} \cdots p_k^{a_k}$$

where $p_1, p_2, \ldots, p_k$ are distinct prime numbers and $a_1, a_2, \ldots, a_k$ are positive integers.

   (a) Use the Fundamental Theorem of Arithmetic to prove that if r and s are natural numbers such that a prime p is a factor of the product rs, then p must be a factor of r or of s (or of both).

   [Hint. Express r and s as products of primes and then use the uniqueness part of the Fundamental Theorem.]

   (b) Let r, s, and b be positive integers such that r and s have no common factors except 1 and -1. Use the Fundamental Theorem of Arithmetic to show that if r is a factor of bs, then r must be a factor of b.

   (c) Let r, s, b, and m be positive integers such that r and s have no common factors except 1 and -1. Show that if r is a factor of bs^m, then r must be a factor of b.

10.* Use the Rational Roots Test to prove that the number $\sin\frac{\pi}{9}$ (that is, $\sin 20°$) is irrational.

   [Hint. First, use the formulae

$$\cos^2\theta + \sin^2\theta = 1,$$
$$\sin(\theta + \phi) = \sin\theta\cos\phi + \cos\theta\sin\phi,$$
$$\cos(\theta + \phi) = \cos\theta\cos\phi - \sin\theta\sin\phi.$$

   to write $\sin 3\theta$ in terms of $\sin\theta$ (without any cos terms). Next, put $\theta = \frac{\pi}{9}$, $x = \sin\theta$, and use $\sin\frac{\pi}{3} = \frac{\sqrt{3}}{2}$.]

Appendix to Chapter 1

We record here formal definitions of ring, field, vector space, and of related terms. If you are not familiar with these concepts, we recommend that you read Sections 23, 24 and 36 of [JF] or Chapters 1 and 2 of [AT] or the early parts of Chapter 3 of [AC].

1.A.1 Definitions. A *group* $(G, +)$ is a set G together with a binary operation $+$ such that

(a) if $g_1, g_2 \in G$, then $g_1 + g_2 \in G$;

(b) if $g_1, g_2, g_3 \in G$, then $g_1 + (g_2 + g_3) = (g_1 + g_2) + g_3$;

(c) there is an element $0 \in G$ such that $g + 0 = 0 + g = g$ for all $g \in G$;

(d) for every element $g \in G$, there exists an element g' such that $g + g' = g' + g = 0$.

The element 0 is called the *identity* of the group G. The element g' in (d) is called the *inverse* of g.

The group G is said to be *abelian* or *commutative* if $g_1 + g_2 = g_2 + g_1$ for all $g_1, g_2 \in G$. ∎

We shall be more interested in sets with two binary operations.

1.A.2 Definitions. A *ring* is a set R with two binary operations $+$ and $.$ such that

(a) R together with the operation $+$ is an abelian group;

(b) if $r_1, r_2 \in R$, then $r_1.r_2 \in R$;

(c) if $r_1, r_2, r_3 \in R$, then $r_1.(r_2.r_3) = (r_1.r_2).r_3$;

(d) if $r_1, r_2, r_3 \in R$, then

$$r_1.(r_2 + r_3) = r_1.r_2 + r_1.r_3 \quad \text{and} \quad (r_2 + r_3).r_1 = r_2.r_1 + r_3.r_1.$$

The additive identity of R is denoted by 0.

The ring R is said to be a *commutative ring* if $r_1.r_2 = r_2.r_1$ for all $r_1, r_2 \in R$.

A *ring with unity* is a ring R with an element 1 (called the *unity* of R) such that $1.r = r.1 = r$ for all $r \in R$.

An element r of a ring $(R, +, .)$ is said to be a *zero-divisor* if $r \neq 0$ and there exists an element $s \neq 0$ in R such that $r.s = 0$ or $s.r = 0$.

Algebraic Preliminaries

A ring $(R, +, .)$ is said to be an *integral domain* if it is a commutative ring with unity and it contains no zero-divisors.

A subset S of a ring $(R, +, .)$ is said to be a *subring* of R if $(S, +, .)$ is itself a ring. ∎

1.A.3 Definitions. A ring $(\mathsf{F}, +, .)$ is said to be a *field* if $\mathsf{F} \setminus \{0\}$ together with the operation . is an abelian group.

A subset S of a field $(\mathsf{F}, +, .)$ is said to be a *subfield* of F if $(S, +, .)$ is itself a field. ∎

The additive identity of a field F is usually denoted by 0, and the multiplicative identity of F is denoted by 1.

Note that while every field is an integral domain, there are integral domains which are not fields: for example, the ring $\mathbb{Z}$ of integers is an integral domain which is not a field.

1.A.4 Definitions. A set V together with two operations + and . together with a field F is said to be a *vector space over the field* F if

(a) V together with the operation + is an abelian group;
(b) if $\lambda \in \mathsf{F}$ and $v \in V$, then $\lambda.v \in V$;
(c) if $\lambda_1, \lambda_2 \in \mathsf{F}$ and $v \in V$, then $(\lambda_1 + \lambda_2).v = \lambda_1.v + \lambda_2.v$;
(d) if $v_1, v_2 \in V$ and $\lambda \in \mathsf{F}$, then $\lambda.(v_1 + v_2) = \lambda.v_1 + \lambda.v_2$;
(e) if 1 is the multiplicative identity of F, then $1.v = v$ for all $v \in V$;
(f) if $\lambda_1, \lambda_2 \in \mathsf{F}$ and $v \in V$, then $(\lambda_1 \lambda_2).v = \lambda_1.(\lambda_2.v)$.

We refer to the elements of F as the *scalars* and the multiplication of an element of F by an element of V (for example, $\lambda.v$ in (b) above) as *scalar multiplication*. The elements of V are called *vectors*. The identity of the group $(V, +)$ is called the *zero vector* and is denoted by 0. The operation + is often called *vector addition*.

A subset S of the vector space V over the field F is said to be a *subspace* of V if S together with the operations + and . of V is a vector space over the field F. ∎

1.A.5 Definitions. If V is a vector space over a field F and S is a subset of V, then the set

$$\text{span}(S) = \{\lambda_1.v_1 + \ldots + \lambda_n.v_n : n \geq 1 \text{ and each } \lambda_i \in \mathsf{F},\ v_i \in S\}$$

is called the *span* of S over F or the *linear span* of S over F.

If $\text{span}(S) = V$ then S is said to *span* V over F. ∎

1.A.6 Definitions. If V is a vector space over a field $\mathbb{F}$ and S is a subset of V, then S is said to be a *linearly independent set over* $\mathbb{F}$ if the zero vector, 0, can be written only as a trivial linear combination of vectors in S; that is, for $v_1, v_2, \ldots, v_n \in S$,

$$0 = \lambda_1.v_1 + \lambda_2.v_2 + \ldots + \lambda_n.v_n$$

implies that $\lambda_1 = \lambda_2 = \ldots = \lambda_n = 0$.

S is said to be *linearly dependent over* $\mathbb{F}$ if it is not linearly independent over $\mathbb{F}$. ∎

1.A.7 Definitions. If V is a vector space over a field $\mathbb{F}$ and S is a nonempty subset of V which is linearly independent over $\mathbb{F}$ and which spans V over $\mathbb{F}$, then S is said to be a *basis of V over* $\mathbb{F}$. If S is finite, then the number of elements of S is called the *dimension of V over* $\mathbb{F}$. ∎

Additional Reading for Chapter 1

General algebra books which include the background information we assume about rings, fields, and vector spaces (and which also deal with geometrical constructions) include [AC], [JF] and [LS], while [AT] is a good introduction to vector spaces. Deeper results about irrational numbers can be found in [GH] (which is the classic reference on number theory) and in [IN1] and [IN2]. [GH] contains a proof of the Fundamental Theorem of Arithmetic and is also a good reference on prime numbers.

[AC] A. Clark, *Elements of Abstract Algebra*, Wadsworth, Belmont, California, 1971.

[JF] J.B. Fraleigh, *A First Course in Abstract Algebra*, 3rd edition, Addison-Wesley, Reading, Massachusetts, 1982.

[GH] G.H. Hardy and E.M. Wright, *An Introduction to the Theory of Numbers*, Clarendon, Oxford, 1960.

[IN1] I. Niven, *Numbers: Rational and Irrational*, Random House, New York, 1961.

[IN2] I. Niven, *Irrational Numbers*, Carus Mathematical Monographs, No. 11, Mathematical Association of America, 1967.

[LS] L.W. Shapiro, *Introduction to Abstract Algebra*, McGraw-Hill, New York, 1975.

[AT] A.M. Tropper, *Linear Algebra*, Thomas Nelson, London, 1969.

CHAPTER 2

Algebraic Numbers

and

Their Polynomials

Straightedge and compass constructions can be used to produce line segments of various lengths relative to some preassigned unit length. Although the lengths are all real numbers, it turns out that not every real number can be obtained in this way. The lengths which can be constructed are rather special.

As the first step towards classifying the lengths which can be constructed, this chapter introduces the concept of an algebraic number (or more specifically of a number which is algebraic over a field). Each such number will satisfy many polynomial equations and our immediate goal is to choose the simplest one.

2.1 Algebraic Numbers

Numbers which lie in $\mathbb{R}$ but not in $\mathbb{Q}$ are said to be *irrational*. Well-known examples are $\sqrt{2}$, e and π. The irrationality of $\sqrt{2}$ was known to the ancient Greeks whereas that of e and π was proved much later. Although these three numbers are all irrational, there is a fundamental distinction between $\sqrt{2}$ and the other two numbers. While $\sqrt{2}$ satisfies a polynomial equation

$$x^2 - 2 = 0$$

with coefficients in $\mathbb{Q}$, no such equation is satisfied by e or π (which we shall prove in Chapter 7.) For this reason $\sqrt{2}$ is said to be an *algebraic number* whereas e and π are said to be *transcendental numbers*.

The precise definition is as follows.

2.1.1 Definition. A number $\alpha \in \mathbb{C}$ is said to be *algebraic over a field* $\mathbb{F} \subseteq \mathbb{C}$ if there is a nonzero polynomial $f(X) \in \mathbb{F}[X]$, such that α is a zero of $f(X)$; that is, there is a polynomial

$$f(X) = a_0 + a_1 X + \ldots + a_n X^n$$

whose coefficients $a_0, a_1, \ldots, a_n$ all belong to $\mathbb{F}$, at least one of these coefficients is nonzero, and $f(\alpha) = 0$. ∎

Observe that for each field $\mathbb{F}$, every number α in $\mathbb{F}$ is algebraic over $\mathbb{F}$ because α is a zero of the polynomial $X - \alpha \in \mathbb{F}[X]$. This implies that e and π are algebraic over $\mathbb{R}$ even though they are not algebraic over $\mathbb{Q}$, as noted above.

2.1.2 Examples.
(i) The number $\sqrt{2}$ is algebraic over $\mathbb{Q}$ because it is a zero of the polynomial $X^2 - 2$, which is nonzero and has coefficients in $\mathbb{Q}$.
(ii) The number $\sqrt[4]{2}\sqrt[6]{3}$ is algebraic over $\mathbb{Q}$ because it is a zero of the polynomial $X^{12} - 72$, which is nonzero and has coefficients in $\mathbb{Q}$. ∎

In order to show that a number is algebraic, we look for a suitable polynomial having that number as a zero.

2.1.3 Example. The number $1 + \sqrt{2}$ is algebraic over $\mathbb{Q}$.

Proof. Let $\alpha = 1 + \sqrt{2}$.

> We want to find a polynomial, with α as a zero, which has coefficients in $\mathbb{Q}$. This suggests squaring to get rid of the square roots:
> $$\alpha^2 = 1 + 2\sqrt{2} + 2,$$
> which is no good as $\sqrt{2}$ still appears on the right hand side. So we go back and isolate $\sqrt{2}$ on one side before squaring!

Isolating $\sqrt{2}$ gives $\qquad \alpha - 1 = \sqrt{2}$.
Squaring both sides gives $\quad (\alpha - 1)^2 = 2$,
and so $\qquad\qquad\qquad\qquad \alpha^2 - 2\alpha - 1 = 0$.

Thus α is a zero of the polynomial $X^2 - 2X - 1$, which is nonzero and has coefficients in $\mathbb{Q}$. Hence α is algebraic over $\mathbb{Q}$. ∎

2.1.4 Other Versions.

It is useful to be able to recognize the definition of "algebraic over a field $\mathbb{F}$" when it appears in different guises: thus $\alpha \in \mathbb{C}$ *is algebraic over $\mathbb{F}$ if and only if*

(i) *there exists $f(X) \in \mathbb{F}[X]$ such that $f(X) \neq 0$ and $f(\alpha) = 0$,*

   OR

(ii) *there is a positive integer n and numbers $a_0, a_1, \ldots, a_{n-1}, a_n$ in $\mathbb{F}$, not all zero, such that*

$$a_0 + a_1\alpha + a_2\alpha^2 + \cdots + a_{n-1}\alpha^{n-1} + a_n\alpha^n = 0.$$

The last statement may ring a few bells! Where have you seen it before? Well, to put things in context recall that, since $\mathbb{F}$ is a subfield of $\mathbb{C}$, *we can regard $\mathbb{C}$ as a vector space over $\mathbb{F}$.* (See the paragraph preceding Definition 1.1.2.) Now the numbers

$$1, \alpha, \alpha^2, \ldots, \alpha^{n-1}, \alpha^n$$

are all elements of $\mathbb{C}$, and hence can be regarded as vectors in the vector space $\mathbb{C}$ over $\mathbb{F}$. The coefficients $a_0, a_1, a_2, \ldots, a_{n-1}, a_n$, on the other hand, are all in $\mathbb{F}$ so we can regard them as scalars. Thus, we can write (ii) in the alternative way:

(iii) there is a positive integer n such that the set of powers

$$\{1, \alpha, \alpha^2, \ldots, \alpha^{n-1}, \alpha^n\}$$

of the number α is `linearly dependent` over F. ∎

You will often meet the terms "algebraic number" and "transcendental number" where no field is specified. In such cases the field is taken to be $\mathbb{Q}$. We formalize this below.

2.1.5 Definitions. A complex number is said to be

(i) an *algebraic number* if it is algebraic over $\mathbb{Q}$;

(ii) a *transcendental number* if it is not algebraic over $\mathbb{Q}$. ∎

——————————— Exercises 2.1 ———————————

1. Verify that each of the following numbers is algebraic over the stated field:
   (a) $\sqrt{2}\,\sqrt[3]{5}$ over $\mathbb{Q}$;
   (b) $\sqrt{2} + \sqrt{3}$ over $\mathbb{Q}$;
   (c) $\sqrt{\pi}$ over $\mathbb{R}$;
   (d) i over $\mathbb{Q}$;
   (e) $\sqrt{2} + \sqrt{3}\,i$ over $\mathbb{R}$.

2. (a) Write down three different polynomials in $\mathbb{Q}[X]$, each of degree 2, which have $\sqrt{5}$ as a zero.
   (b) Write down three different polynomials in $\mathbb{Q}[X]$, each of degree greater than 2, which have $\sqrt{5}$ as a zero.
   (c) Write down three different polynomials in $\mathbb{Q}[X]$, each of degree 100, which have $\sqrt{5}$ as a zero.

3. Prove that if $\alpha \in \mathbb{C}$ is algebraic over $\mathbb{Q}$ then so is 2α.

4. Prove that if $\alpha \in \mathbb{C}$ is algebraic over a subfield F of $\mathbb{C}$ then so is $-\alpha$.

5. Prove that if α is a nonzero complex number which is algebraic over a subfield F of $\mathbb{C}$, then $1/\alpha$ is also algebraic over F.

Algebraic Numbers

6. Show that every complex number is algebraic over $\mathbb{R}$.

7.* Let α be a positive real number, and let $\mathbb{F}$ be a subfield of $\mathbb{R}$. Prove that α is algebraic over $\mathbb{F}$ if and only if $\sqrt{\alpha}$ is algebraic over $\mathbb{F}$.

2.2 Monic Polynomials

An algebraic number such as $\sqrt{2}$ will be a zero of many different polynomials. Ultimately we want to pick out from all these polynomials one which is, in some sense, the best. Here we make a start in this direction by restricting attention to monic polynomials, defined as follows:

2.2.1 Definition. A polynomial $f(X) = a_0 + a_1 X + \cdots + a_n X^n$ in $\mathbb{F}[X]$ is said to be *monic* if its leading coefficient a_n is 1. ∎

Thus, for example, $X^2 - 2$ is a monic polynomial whereas $3X^2 - 6$ is not. Note that both these polynomials have $\sqrt{2}$ as a zero and, for our purposes, the polynomial $X^2 - 2$ is somewhat nicer than $3X^2 - 6$. The following proposition shows that, in the definition of "algebraic over a field", we can always assume the polynomial is monic.

2.2.2 Proposition. *If a complex number α is a zero of a nonzero polynomial $f(X) \in \mathbb{F}[X]$ then α is a zero of a monic polynomial $g(X) \in \mathbb{F}[X]$ with $\deg g(X) = \deg f(X)$.*

Proof. Assume that α is a zero of a polynomial $f(X) \neq 0$. Since $f(X) \neq 0$, some coefficient must be nonzero and (because there are only finitely many coefficients a_i) there must be a largest i such that $a_i \neq 0$. Let n be the largest such i.

Hence
$$f(X) = a_0 + a_1 X + a_2 X^2 + \ldots + a_n X^n$$
where $a_n \neq 0$. We choose
$$g(X) = \frac{1}{a_n} f(X).$$
Thus $g(X)$ is monic, has α as a zero and the same degree n as $f(X)$. All the coefficients of $g(X)$ are in $\mathbb{F}$, moreover, since $\mathbb{F}$ is a field. ∎

_____ Exercises 2.2 _____

1. If α is a zero of $3X^3 - 2X + 1$, find a monic polynomial with coefficients in $\mathbb{Q}$ having α as a zero.

2. If $2\alpha^3 - 1 = \sqrt{3}$, find a monic polynomial in $\mathbb{Q}[X]$ which has α as a zero.

3. (a) Write down three different monic polynomials in $\mathbb{Q}[X]$ which have $\sqrt[3]{5}$ as a zero.
   (b) Write down three different monic polynomials in $\mathbb{Q}[X]$, each of degree 100, which have $\sqrt[3]{5}$ as a zero.
   (c) How many monic polynomials in $\mathbb{Q}[X]$ have $\sqrt[3]{5}$ as a zero?

2.3 Monic Polynomials of Least Degree

Even if we restrict attention to monic polynomials, there are still a lot of them which have $\sqrt{2}$ as a zero. For example, $\sqrt{2}$ is a zero of each of the polynomials

$$X^2 - 2, \quad X^4 - 4, \quad (X^2 - 2)^2, \quad (X^2 - 2)(X^5 + 3),$$
$$(X^2 - 2)(X^{100} + 84X^3 + 73)$$

all of which are in $\mathbb{Q}[X]$. What distinguishes $X^2 - 2$ from the other polynomials is that it has the least degree.

If a number α is a zero of one monic polynomial then (as suggested by Exercises 2.2 #3) there are infinitely many monic polynomials of arbitrarily high degree having α as a zero. Although there is no maximum possible degree for such polynomials, there is always a least possible degree. (For example, if we know α is a zero of a monic polynomial of degree 12, we can be sure that the least possible degree is one of the finitely-many numbers $1, 2, 3, \ldots, 11, 12$.)

The following proposition shows that by focusing on the least possible degree, we pick out a unique polynomial.

2.3.1 Proposition. *If $\alpha \in \mathbb{C}$ is algebraic over a field $\mathbb{F} \subseteq \mathbb{C}$ then among all monic polynomials $f(X) \in \mathbb{F}[X]$ with $f(\alpha) = 0$ there is a unique one of least degree.*

Algebraic Numbers

Proof. Since α is algebraic, there is (by Proposition 2.2.2) a monic polynomial $f(X) \in \mathsf{F}[X]$ with $f(\alpha) = 0$. Hence there is one such polynomial of least possible degree.

To prove there is <u>only one</u> such polynomial, suppose there are two of them, say $f_1(X)$ and $f_2(X)$, each having the least degree n, say. Thus
$$f_1(X) = a_0 + a_1 X + \cdots + a_{n-1} X^{n-1} + X^n$$
$$f_2(X) = b_0 + b_1 X + \cdots + b_{n-1} X^{n-1} + X^n.$$
Hence if we put $f(X) = f_1(X) - f_2(X)$ we get
$$f(X) = c_0 + c_1 X + \ldots + c_{n-1} X^{n-1}, \quad \text{where } c_i = a_i - b_i.$$

⌐ We want to prove that $f_1(X) = f_2(X)$. To do this we show that $f(X) = 0$. ⌐

Clearly $f(X) \in \mathsf{F}[X]$, $f(\alpha) = 0$ and either $f(X)$ is zero or $f(X)$ has degree $< n$. If $f(X) \neq 0$ then, by Proposition 2.2.2, there is a monic polynomial which has the same degree as $f(X)$ and which has α as a zero. But this is impossible since n is the least degree for which there is a monic polynomial having α as a zero.

It follows that $f(X) = 0$, and hence $f_1(X) = f_2(X)$. ∎

The above theorem enables us to assign to each algebraic number a unique polynomial. This polynomial is so important that we introduce some terminology and notation for it.

2.3.2 Definitions. Let $\alpha \in \mathbb{C}$ be algebraic over a field $\mathsf{F} \subseteq \mathbb{C}$. The unique polynomial of least degree among those polynomials $f(X)$ in $\mathsf{F}[X]$ satisfying

(i) $f(\alpha) = 0$ and (ii) $f(X)$ is monic

is called *the irreducible polynomial of α over* F. This polynomial is denoted by
$$\mathrm{irr}(\alpha, \mathsf{F}).$$
Its degree is called the *degree of α over* F and is denoted by
$$\deg(\alpha, \mathsf{F}). \qquad \blacksquare$$

The word "irreducible" means "cannot be reduced" and its use in the above definition stems from the idea that we cannot reduce the degree below that of irr(α, F) if we still want to have a polynomial with the desired properties (i) and (ii).

Later you will find that irr(α, F) is also irreducible in a different sense. Thus the use of the word "irreducible" is doubly justified.

2.3.3 Example. We prove that the irreducible polynomial of $\sqrt{2}$ over $\mathbb{Q}$ is $X^2 - 2$.

Proof. It is clear that this polynomial has $\sqrt{2}$ as a zero, that its coefficients are in $\mathbb{Q}$ and that it is monic. It remains to show there is no polynomial of smaller degree with these properties.

If there were such a polynomial it would be

$$a + X, \quad \text{for some } a \in \mathbb{Q}.$$

But then we would have

$$a + \sqrt{2} = 0$$

so that $\sqrt{2} = -a \in \mathbb{Q}$, which is a contradiction.

Hence

$$\text{irr}(\sqrt{2}, \mathbb{Q}) = X^2 - 2$$

and therefore

$$\deg(\sqrt{2}, \mathbb{Q}) = 2. \qquad \blacksquare$$

Note that, although the irreducible polynomial of $\sqrt{2}$ over $\mathbb{Q}$ is $X^2 - 2$, its irreducible polynomial over $\mathbb{R}$ is $X - \sqrt{2}$. This is because $X - \sqrt{2}$ is in $\mathbb{R}[X]$ but not in $\mathbb{Q}[X]$.

The idea in the worked example above is quite simple – find the polynomial you think is of least degree and, to show it really is of least degree, consider all smaller degrees. This can be far from straightforward if the polynomial in question has degree greater than 2. For example, you would suspect that irr$(\sqrt[7]{2}, \mathbb{Q})$ is $X^7 - 2$. But to prove that $\sqrt[7]{2}$ is not a zero of any monic polynomial in $\mathbb{Q}[X]$ of degrees 1, 2, 3, 4, 5 or 6 would be very difficult (to say the least) with the techniques you know at present. Special techniques have been developed for these sorts of problems; you will meet some of them in Chapter 4.

Algebraic Numbers

---------- **Exercises 2.3** ----------

1. (a) Write down the irreducible polynomial of $\sqrt{5}$ over $\mathbb{Q}$ and then prove your answer is correct.
   (b) Write down $\deg(\sqrt{5}, \mathbb{Q})$.

2. In each case write down a nonzero polynomial $f(X)$ satisfying the stated conditions:
   (a) $f(X)$ is a monic polynomial over $\mathbb{Q}$ with $\sqrt{2}$ as a zero.
   (b) $f(X)$ is a polynomial over $\mathbb{Q}$ with $\sqrt{2}$ as a zero but is not monic.
   (c) $f(X)$ is a polynomial of least degree such that
   $$f(X) \in \mathbb{R}[X] \quad \text{and} \quad f(\sqrt{2}) = 0.$$
   (d) $f(X)$ is another polynomial satisfying conditions (c).
   (e) $f(X) \in \mathbb{Q}[X]$ and has both $\sqrt{2}$ and $\sqrt{3}$ as zeros.

3. (a) In each case write down the unique monic polynomial $f(X)$ of least degree which has the stated property:
   (i) $f(X) \in \mathbb{R}[X]$ and has $\sqrt{3}$ as a zero.
   (ii) $f(X) \in \mathbb{Q}[X]$ and has $\sqrt{3}$ as a zero.
   (b) Explain why the polynomials you gave in part (a) have least degree.
   (c) What do your answers to part (a) tell you about
   (i) $\mathrm{irr}(\sqrt{3}, \mathbb{R})$ and $\deg(\sqrt{3}, \mathbb{R})$?
   (ii) $\mathrm{irr}(\sqrt{3}, \mathbb{Q})$ and $\deg(\sqrt{3}, \mathbb{Q})$?

4. Assume α is algebraic over $\mathbb{F}$. Let n be the least degree of all monic nonzero polynomials $f(X)$ in $\mathbb{F}[X]$ with $f(\alpha) = 0$ and let m be the least degree of all nonzero polynomials $g(X)$ (monic or not) in $\mathbb{F}[X]$ with $g(\alpha) = 0$. Explain why $m = n$. (Which theorem or proposition from an earlier section do you use?)

5. Let $\mathbb{F}$ be a subfield of $\mathbb{R}$ and let $a, b, c \in \mathbb{F}$ with c positive.
   (a) Show that $a + b\sqrt{c}$ is a zero of a monic quadratic polynomial in $\mathbb{F}[X]$.
   (b) What can be said about the degree of $a + b\sqrt{c}$ over $\mathbb{F}$?

6. In accordance with # 4, 5, 7 of Exercises 2.1, if $\alpha \in \mathbb{R}$ is algebraic over $\mathbb{F}$ then so are $-\alpha$, $\sqrt{\alpha}$ when $\alpha > 0$, and $1/\alpha$ when $\alpha \neq 0$.
   (a) What is the relationship between $\deg(\alpha, \mathbb{F})$ and $\deg(-\alpha, \mathbb{F})$?
   (b) Prove that $\deg(\sqrt{\alpha}, \mathbb{F}) \leq 2 \deg(\alpha, \mathbb{F})$.
   (c) Is the following statement true or false? (Justify your answer.) If $\alpha \neq 0$ is algebraic over $\mathbb{F}$ then
   $$\deg(\alpha, \mathbb{F}) = \deg\left(\frac{1}{\alpha}, \mathbb{F}\right).$$

7. Let $\mathbb{F} \subseteq \mathbb{C}$ and let $\alpha \in \mathbb{C}$. In each case decide whether the statement is true or false (and justify your answer).
   (a) If $\deg(\alpha, \mathbb{F}) = 2$ then $\alpha \notin \mathbb{F}$.
   (b) If $\deg(\alpha, \mathbb{F}) = 2$ then $\alpha^2 \in \mathbb{F}$.

8. If $\alpha \in \mathbb{C}$ is algebraic over a subfield $\mathbb{F}$ of $\mathbb{C}$ and
   $$\{1, \alpha, \alpha^2, \alpha^3\}$$
   is linearly independent over $\mathbb{F}$, what can be said about $\deg(\alpha, \mathbb{F})$?

9. Let $\mathbb{F}$ be a subfield of $\mathbb{C}$ and let α be a complex number. Assume that the set $\{1, \alpha, \alpha^2 \alpha^3, \alpha^4\}$ is linearly dependent over $\mathbb{F}$.
   (a) Prove that α is algebraic over $\mathbb{F}$.
   (b) What can be said about $\deg(\alpha, \mathbb{F})$? (Justify your answer.)

10. Let $\mathbb{F}$ be a subfield of $\mathbb{C}$ and let $\alpha \in \mathbb{C}$. Prove that $\deg(\alpha, \mathbb{F}) = 1$ if and only if $\alpha \in \mathbb{F}$.

11. Is the following statement true or false? (Justify your answer.) If α, β and $\alpha\beta$ are algebraic over $\mathbb{Q}$ then
    $$\deg(\alpha\beta, \mathbb{Q}) = \deg(\alpha, \mathbb{Q}) \deg(\beta, \mathbb{Q}).$$

12. Let $\mathbb{E} \subseteq \mathbb{F}$ be subfields of $\mathbb{C}$ and assume that $\beta \in \mathbb{C}$ is algebraic over $\mathbb{E}$.
    (a) Prove that β is algebraic over $\mathbb{F}$ also.
    (b) Prove that $\deg(\beta, \mathbb{E}) \geq \deg(\beta, \mathbb{F})$.

13. Let $\mathbb{E} \subseteq \mathbb{F}$ be subfields of $\mathbb{C}$ such that $[\mathbb{F} : \mathbb{E}] = 5$. Let $\beta \in \mathbb{F}$.
    (a) What is the maximum number of elements which can be in a subset of $\mathbb{F}$ which is linearly independent over $\mathbb{E}$?

(b) State why the following powers of β are all in $\mathbb{F}$:
$$1, \beta, \beta^2, \beta^3, \beta^4, \beta^5.$$

(c) Deduce from (a) and (b) that β is algebraic over $\mathbb{E}$.

(d) What can be said about (i) $\deg(\beta, \mathbb{E})$? (ii) $\deg(\beta, \mathbb{F})$?

14. Let $\mathbb{F}$ be a subfield of $\mathbb{C}$ and let $\alpha \in \mathbb{C}$. Let $f(X) \in \mathbb{F}[X]$ be a monic polynomial with $f(\alpha) = 0$. If there are polynomials $g(X), h(X)$ in $\mathbb{F}[X]$, each of degree ≥ 1, such that $f(X) = g(X) h(X)$, show that $f(X)$ is not $\mathrm{irr}(\alpha, \mathbb{F})$.

Additional Reading for Chapter 2

We have introduced $\mathrm{irr}(\alpha, \mathbb{F})$ with very few preliminary results about polynomials. However, treatments of $\mathrm{irr}(\alpha, \mathbb{F})$ in most other books depend heavily on techniques for showing that a particular polynomial cannot be reduced (that is, factorized further). For this reason, you may find it confusing to read about $\mathrm{irr}(\alpha, \mathbb{F})$ in other books before you have finished reading our Chapter 4.

A discussion of numbers algebraic over a field is given in the first two sections of Chapter XIV of [GB]. For some historical facts about transcendental numbers see Section 5.4 of [IN1].

[GB] G. Birkhoff and S. MacLane, *A Survey of Modern Algebra*, New York, Macmillan, 1953.

[IN1] I. Niven, *Numbers: Rational and Irrational*, Random House, New York, 1961.

CHAPTER 3

Extending Fields

If $\mathbb{F}$ is a subfield of $\mathbb{C}$ and α is a complex number which is algebraic over $\mathbb{F}$, we show how to construct a certain vector space $\mathbb{F}(\alpha)$ which contains α and which satisfies

$$\mathbb{F} \subseteq \mathbb{F}(\alpha) \subseteq \mathbb{C}.$$

This vector space is then shown to be a subfield of $\mathbb{C}$. Thus, from the field $\mathbb{F}$ and the number α, we have produced a larger field $\mathbb{F}(\alpha)$.

Fields of the form $\mathbb{F}(\alpha)$ are essential to our analysis of the lengths of those line segments which can be constructed with straightedge and compass.

3.1 An Illustration: $\mathbb{Q}(\sqrt{2})$

As $\mathbb{Q}$ is a subfield of $\mathbb{C}$, we can consider $\mathbb{C}$ as a vector space over $\mathbb{Q}$, taking the elements of $\mathbb{C}$ as the vectors and the elements of $\mathbb{Q}$ as the scalars.

3.1.1 Definition. The set $\mathbb{Q}(\sqrt{2}) \subseteq \mathbb{C}$ is defined by putting

$$\mathbb{Q}(\sqrt{2}) = \{a + b\sqrt{2} : a, b \in \mathbb{Q}\}. \qquad \blacksquare$$

Thus $\mathbb{Q}(\sqrt{2})$ is the *linear span* of the set of vectors $\{1, \sqrt{2}\}$ over $\mathbb{Q}$ and is therefore a vector *subspace* of $\mathbb{C}$ over $\mathbb{Q}$. Hence $\mathbb{Q}(\sqrt{2})$ is a vector space over $\mathbb{Q}$.

3.1.2 Proposition. *The set of vectors $\{1, \sqrt{2}\}$ is a basis for the vector space $\mathbb{Q}(\sqrt{2})$ over $\mathbb{Q}$.*

Proof. As noted above, this set of vectors spans $\mathbb{Q}(\sqrt{2})$.

Hence it is sufficient to establish its linear independence over $\mathbb{Q}$. To this end, let $a, b \in \mathbb{Q}$ with

$$a + b\sqrt{2} = 0. \qquad (1)$$

If $b \neq 0$, then $\sqrt{2} = -a/b$, which is again in $\mathbb{Q}$ as $\mathbb{Q}$ is a field. This contradicts the fact that $\sqrt{2}$ is irrational. Hence $b = 0$. It now follows from (1) that also $a = 0$.

Thus (1) implies $a = 0$ and $b = 0$ so that the set of vectors $\{1, \sqrt{2}\}$ is linearly independent over $\mathbb{Q}$. $\blacksquare$

Note that the dimension of the vector space $\mathbb{Q}(\sqrt{2})$ over $\mathbb{Q}$ is 2, this being the number of vectors in the basis

3.1.3 Proposition. $\mathbb{Q}(\sqrt{2})$ *is a ring.*

Proof. To show that $\mathbb{Q}(\sqrt{2})$ is a subring of $\mathbb{C}$, it is sufficient to check that it is closed under addition and multiplication. The former is obvious while the latter is left as a simple exercise. $\blacksquare$

3.1.4 Proposition. $\mathbb{Q}(\sqrt{2})$ *is a field.*

Proof. To show that this subring of $\mathbb{C}$ is a subfield, it is sufficient to check that it contains the reciprocal of each of its nonzero elements.

So let $x \in \mathbb{Q}(\sqrt{2})$ be such that $x \neq 0$. Thus
$$x = a + b\sqrt{2}$$
where $a, b \in \mathbb{Q}$ and $a \neq 0$ or $b \neq 0$. It follows that
$$a - b\sqrt{2} \neq 0$$
by the linear independence of the set $\{1, \sqrt{2}\}$. Thus
$$\begin{aligned}
\frac{1}{x} &= \frac{1}{a + b\sqrt{2}} \\
&= \frac{1}{a + b\sqrt{2}} \cdot \frac{a - b\sqrt{2}}{a - b\sqrt{2}} \\
&= \left(\frac{a}{a^2 - 2b^2}\right) + \left(\frac{-b}{a^2 - 2b^2}\right)\sqrt{2}
\end{aligned}$$
which is again an element of $\mathbb{Q}(\sqrt{2})$, since $a/(a^2-2b^2)$ and $-b/(a^2-2b^2)$ are both in $\mathbb{Q}$. ∎

The following proposition gives a way of describing $\mathbb{Q}(\sqrt{2})$ as a field with a certain property. The proof, like that of Proposition 1.1.1, involves proving a set inclusion.

3.1.5 Proposition. $\mathbb{Q}(\sqrt{2})$ *is the smallest field containing all the numbers in the field* $\mathbb{Q}$ *and the number* $\sqrt{2}$.

Proof. Let F be any field containing $\mathbb{Q}$ and $\sqrt{2}$.

> It is clear from what has been said earlier in this section that $\mathbb{Q}(\sqrt{2})$ is a field which contains both $\mathbb{Q}$ and $\sqrt{2}$. To show it is the smallest such field we shall prove that
> $$\mathbb{Q}(\sqrt{2}) \subseteq \mathsf{F}.$$

To prove that $\mathbb{Q}(\sqrt{2}) \subseteq \mathsf{F}$ we shall show that
$$\text{if } x \in \mathbb{Q}(\sqrt{2}), \text{ then } x \in \mathsf{F}. \tag{2}$$
To prove this we start by assuming the hypothesis of (2). Let $x \in \mathbb{Q}(\sqrt{2})$; that is,
$$x = a + b\sqrt{2} \qquad \text{for some } a, b \in \mathbb{Q}.$$

> Our aim is to prove that $x \in \mathsf{F}$. To do this we shall use the fact that F is a field.

By assumption, $\sqrt{2} \in \mathsf{F}$.
Also $a, b \in \mathsf{F}$ as F is assumed to contain $\mathbb{Q}$.
Hence $b\sqrt{2} \in \mathsf{F}$ as F is closed under multiplication.
So $a + b\sqrt{2} \in \mathsf{F}$ as F is closed under addition.
Thus $x \in \mathsf{F}$, as required. ∎

Because $\mathbb{Q}(\sqrt{2})$ is a field, the theory developed in Chapter 2 for a field $\mathsf{F} \subseteq \mathbb{C}$ can now be applied in the case $\mathsf{F} = \mathbb{Q}(\sqrt{2})$. The following example illustrates this.

3.1.6 Example. The number $\sqrt{3}$ is algebraic over $\mathbb{Q}(\sqrt{2})$ and has degree 2 over this field.

Proof. The polynomial $X^2 - 3$ has coefficients in $\mathbb{Q}$, and hence also in $\mathbb{Q}(\sqrt{2})$. It is monic, moreover, and has $\sqrt{3}$ as a zero. Hence $\sqrt{3}$ is algebraic over $\mathbb{Q}(\sqrt{2})$.

Suppose $X^2 - 3$ is not the irreducible polynomial of $\sqrt{3}$ over $\mathbb{Q}(\sqrt{2})$. Then the irreducible polynomial is a monic first degree polynomial

$$a + X$$

for some $a \in \mathbb{Q}(\sqrt{2})$. Hence $a + \sqrt{3} = 0$ and so $\sqrt{3} = -a$, which implies that

$$\sqrt{3} = c + d\sqrt{2}$$

for some $c, d \in \mathbb{Q}$. Squaring both sides gives

$$3 = c^2 + 2\sqrt{2}cd + 2d^2.$$

If

$$c = 0$$

we have $2d^2 = 3$ which leads to the contradiction that $\sqrt{3/2}$ is rational, while if

$$d = 0$$

we get the contradiction that $\sqrt{3}$ is rational. Hence $cd \neq 0$, which gives

$$\sqrt{2} = \frac{3 - c^2 - 2d^2}{2cd} \in \mathbb{Q},$$

Extending Fields

which is also a contradiction.

Thus our original assumption that $X^2 - 3$ is not the irreducible polynomial of $\sqrt{3}$ over $\mathbb{Q}(\sqrt{2})$ has led to a contradiction. Hence $X^2 - 3$ is the irreducible polynomial of $\sqrt{3}$ over $\mathbb{Q}(\sqrt{2})$; that is,

$$\mathrm{irr}(\sqrt{3},\ \mathbb{Q}(\sqrt{2})) = X^2 - 3$$

and hence

$$\deg(\sqrt{3},\ \mathbb{Q}(\sqrt{2})) = 2. \qquad \blacksquare$$

─────────────── **Exercises 3.1** ───────────────

1. Verify that each of the following numbers is in $\mathbb{Q}(\sqrt{2})$ by expressing it in the form $a + b\sqrt{2}$, where $a, b \in \mathbb{Q}$.
   (i) $(2 + 3\sqrt{2})(1 - 2\sqrt{2})$,
   (ii) $(1 + 3\sqrt{2})/(3 + 2\sqrt{2})$,
   (iii) $\sqrt{3 + 2\sqrt{2}}$. [Hint. Let this equal $a + b\sqrt{2}$ and square both sides.]

2. Show that the number $\sqrt{1 + \sqrt{2}}$ is not in $\mathbb{Q}(\sqrt{2})$.
   [Hint. Let this number equal $a + b\sqrt{2}$, square both sides and derive a contradiction.]

3. Show that the zeros in $\mathbb{C}$ of the polynomial $X^2 + 2X - 7$ are both in $\mathbb{Q}(\sqrt{2})$.

4. Verify that the number $\sqrt{5}$ is algebraic over $\mathbb{Q}(\sqrt{2})$ and find its degree over this field.

5. Verify that the number $\sqrt{144}$ is algebraic over $\mathbb{Q}(\sqrt{2})$ and find its degree over this field.

6. Complete the proof of Proposition 3.1.3, begun in the text.
   [Hint. Assume that $x, y \in \mathbb{Q}(\sqrt{2})$ and then spell out explicitly what this means. Hence show that $xy \in \mathbb{Q}(\sqrt{2})$.]

7. Verify that $\sqrt{m}$ is algebraic over each of the fields $\mathbb{Q}$ and $\mathbb{Q}(\sqrt{2})$, for each integer $m \geq 0$. State in which cases it is true that
   $$\deg(\sqrt{m},\ \mathbb{Q}) = \deg(\sqrt{m},\ \mathbb{Q}(\sqrt{2})).$$

8. Let $a, b \in \mathbb{Q}$. Show that if $a + b\sqrt{2}$ is a zero of a quadratic polynomial in $\mathbb{Q}[X]$ then its "conjugate" $a - b\sqrt{2}$ is also a zero of this polynomial.

9.* Prove the result of Exercise 8, but without the restriction that the polynomial is quadratic.

3.2 Construction of $\mathbb{F}(\alpha)$

In the previous section we constructed, from the field $\mathbb{Q}$ and the number $\sqrt{2}$, a vector subspace $\mathbb{Q}(\sqrt{2})$ of $\mathbb{C}$ by putting

$$\mathbb{Q}(\sqrt{2}) = \{a + b\sqrt{2} : a, b \in \mathbb{Q}\}.$$

We then found that $\mathbb{Q}(\sqrt{2})$ is closed under multiplication (as well as addition) and hence is a ring. Next $\mathbb{Q}(\sqrt{2})$ was shown to contain the reciprocal of each of its nonzero elements, and hence it is also a field. Finally we showed that $\mathbb{Q}(\sqrt{2})$ was the *smallest subfield of* $\mathbb{C}$ *containing all the numbers in* $\mathbb{Q}$ *together with the number* $\sqrt{2}$.

In this section we generalize the construction of $\mathbb{Q}(\sqrt{2})$ by constructing a subset $\mathbb{F}(\alpha)$ of $\mathbb{C}$ for any subfield $\mathbb{F}$ of $\mathbb{C}$ and any number $\alpha \in \mathbb{C}$ which is algebraic over $\mathbb{F}$. We begin by defining $\mathbb{F}(\alpha)$ and then showing that this set is in turn a vector subspace, a subring, and then a subfield of $\mathbb{C}$. Our final goal is Theorem 3.2.8, which says that $\mathbb{F}(\alpha)$ is *the smallest subfield of* $\mathbb{C}$ *containing* α *and all the numbers in* $\mathbb{F}$. Throughout this section, we assume that α is algebraic over $\mathbb{F}$.

3.2.1 Definition. Let $\mathbb{F}$ be a subfield of $\mathbb{C}$ and let $\alpha \in \mathbb{C}$ be algebraic over $\mathbb{F}$ with $\deg(\alpha, \mathbb{F}) = n$. The *extension of* $\mathbb{F}$ *by* α is the set $\mathbb{F}(\alpha) \subseteq \mathbb{C}$ where

$$\mathbb{F}(\alpha) = \{b_0 + b_1\alpha + \ldots + b_{n-1}\alpha^{n-1} : b_0, b_1, \ldots, b_{n-1} \in \mathbb{F}\}. \quad \blacksquare$$

Thus $\mathbb{F}(\alpha)$ is the linear span over $\mathbb{F}$ of the powers

$$1, \alpha, \alpha^2, \ldots, \alpha^{n-1}$$

and so is a *vector subspace* of $\mathbb{C}$ over $\mathbb{F}$. In this section we show that $\mathbb{F}(\alpha)$ is a field – indeed it is the smallest field containing $\mathbb{F}$ and α. These results are proved in Theorems 3.2.6 and 3.2.8.

Why did we stop at the power α^{n-1}? The answer is given by the following result, which shows that any further powers would be redundant.

3.2.2 Proposition. *The set $\mathbb{F}(\alpha)$ contains all the remaining positive powers of α:*
$$\alpha^n, \alpha^{n+1}, \alpha^{n+2}, \alpha^{n+3}, \ldots .$$

Proof. Since $n = \deg(\alpha, \mathbb{F})$, α is a zero of a monic polynomial of degree n in $\mathbb{F}[X]$. Hence
$$c_0 + c_1\alpha + \ldots + c_{n-1}\alpha^{n-1} + \alpha^n = 0$$
for some coefficients $c_0, c_1, \ldots, c_{n-1} \in \mathbb{F}$. Thus
$$\alpha^n = -c_0 - c_1\alpha - \ldots - c_{n-1}\alpha^{n-1} \tag{1}$$
and each $-c_i \in \mathbb{F}$, as $\mathbb{F}$ is a field. Hence, by the definition of $\mathbb{F}(\alpha)$ (Definition 3.2.1),
$$\alpha^n \in \mathbb{F}(\alpha).$$
If we multiply both sides of (1) by α we see that
$$\alpha^{n+1} = -c_0\alpha - \ldots - c_{n-2}\alpha^{n-1} - c_{n-1}\alpha^n. \tag{2}$$
Now $\alpha, \ldots, \alpha^{n-1}$ are all in $\mathbb{F}(\alpha)$ (by definition) and we have just shown that $\alpha^n \in \mathbb{F}(\alpha)$. Thus, because $\mathbb{F}(\alpha)$ is a vector subspace of $\mathbb{C}$, it follows from (2) that
$$\alpha^{n+1} \in \mathbb{F}(\alpha).$$
Now multiply both sides of (1) by α^2 and then proceed in the same way, thereby showing that
$$\alpha^{n+2} \in \mathbb{F}(\alpha).$$
It is clear that by proceeding in this way we can show that the powers
$$\alpha^n, \alpha^{n+1}, \alpha^{n+2}, \ldots$$
all belong to $\mathbb{F}(\alpha)$. ∎

Thus, in our definition of $\mathbb{F}(\alpha)$ we have, in some sense, included "enough" powers of α. Have we included "too many"? An answer can be deduced from the following result.

3.2.3 Proposition. *If n is the degree of α over F, then the set of vectors $\{1, \alpha, \alpha^2, \ldots, \alpha^{n-1}\}$ is linearly independent over F.*

Proof. Suppose on the contrary that this set is linearly dependent; so there are scalars $c_0, c_1, \ldots, c_{n-1} \in \mathsf{F}$, not all zero, such that

$$c_0 + c_1\alpha + \ldots + c_{n-1}\alpha^{n-1} = 0.$$

Among the coefficients $c_0, c_1, \ldots, c_{n-1}$ pick the one, say c_i, farthest down the list which is nonzero. Dividing by this coefficient gives

$$\left(\frac{c_0}{c_i}\right) + \left(\frac{c_1}{c_i}\right)\alpha + \ldots + \left(\frac{c_{i-1}}{c_i}\right)\alpha^{i-1} + \alpha^i = 0$$

where $i \leq n - 1$, and the coefficients are still in the field F.

Hence α is a zero of a monic polynomial in $\mathsf{F}[X]$ with degree smaller than n (which was the least possible degree for such polynomials). This contradiction shows that our initial assumption was false. ∎

Suppose we had tried to take

$$S = \{b_0 + b_1\alpha + \ldots + b_{n-2}\alpha^{n-2} : b_0, \ldots, b_{n-2} \in \mathsf{F}\}$$

as a candidate for a ring (or field) containing F and α. Then (if $n \geq 2$), α and α^{n-2} are both in S but their product α^{n-1} is not in S by the above proposition. Thus S is too small and so we need all the powers $1, \alpha, \ldots, \alpha^{n-1}$ in $\mathsf{F}(\alpha)$ for it to be a ring or a field.

3.2.4 Theorem. [Basis for $\mathsf{F}(\alpha)$ Theorem] *Let F be a subfield of $\mathbb{C}$ and let $\alpha \in \mathbb{C}$ be algebraic over F with $\deg(\alpha, \mathsf{F}) = n$. Then the set of vectors $\{1, \alpha, \alpha^2, \ldots, \alpha^{n-1}\}$ is a basis for the vector space $\mathsf{F}(\alpha)$ over F. In particular this vector space has dimension n, the degree of α over F.*

Proof. By Definition 3.2.1, this set of vectors spans the vector space $\mathsf{F}(\alpha)$ over F and, by Proposition 3.2.3, the set of vectors is linearly independent. Hence it is a basis for the vector space and the number n of elements in the basis is the dimension of the vector space. ∎

As $\mathsf{F}(\alpha)$ is a subset of $\mathbb{C}$, its elements can be multiplied together. This leads to the following proposition.

Extending Fields

3.2.5 Proposition. $F(\alpha)$ *is a ring.*

Proof. To show $F(\alpha)$ is a subring of $\mathbb{C}$, it is sufficient to show it is closed under addition and multiplication. The former is obvious while the latter will now be shown. Consider any two elements in $F(\alpha)$, say
$$p = b_0 + b_1\alpha + b_2\alpha^2 + \ldots + b_{n-1}\alpha^{n-1} \in F(\alpha)$$
and
$$q = c_0 + c_1\alpha + c_2\alpha^2 + \ldots + c_{n-1}\alpha^{n-1} \in F(\alpha),$$
where the b's and c's belong to F. If we multiply these two elements together, we get a linear combination of powers of α, with coefficients still in the field F. Because all positive powers of α are in $F(\alpha)$ (by Proposition 3.2.2) and because $F(\alpha)$ is closed under addition and scalar multiplication, it follows that the product pq is also in $F(\alpha)$. ∎

At long last we are in a position to establish that $F(\alpha)$ is a field.

3.2.6 Theorem. *Let F be a subfield of $\mathbb{C}$ and let $\alpha \in \mathbb{C}$ be algebraic over F. Then $F(\alpha)$ is a field.*

Proof. In view of Proposition 3.2.5, what remains to be shown is that $1/\beta$ is in $F(\alpha)$ for every nonzero β in $F(\alpha)$. So let β be a nonzero number in $F(\alpha)$.

Firstly note that $\{1, \beta, \beta^2, \ldots, \beta^n\}$ is a set of $n+1$ numbers which are all in $F(\alpha)$, since $F(\alpha)$ is closed under multiplication. Because $F(\alpha)$ is an n–dimensional vector space over F, this set must be linearly dependent, which means that there are scalars $d_0, d_1, \ldots, d_k$ in F (not all zero) with $k \leq n$ such that
$$d_0 + d_1\beta + d_2\beta^2 + \ldots + d_k\beta^k = 0. \tag{3}$$
If $d_0 = 0$ in (3), we could divide by β (or multiply by $1/\beta$) to reduce the number of terms in (3). By repeating this if necessary, we see that (3) can be assumed to hold with $d_0 \neq 0$. If we multiply (3) by $-1/d_0$ we have
$$-1 + e_1\beta + e_2\beta^2 + \ldots + e_k\beta^k = 0,$$
where each $e_i = -d_i/d_0$ is in F since F is a field. Thus
$$1 = e_1\beta + e_2\beta^2 + \ldots + e_k\beta^k = \beta(e_1 + e_2\beta + \ldots + e_k\beta^{k-1}).$$
It follows that
$$1/\beta = e_1 + e_2\beta + \ldots + e_k\beta^{k-1},$$
which is in $F(\alpha)$ since the e_i and β^i are in $F(\alpha)$ and $F(\alpha)$ is a ring. ∎

In simple cases the method used to prove Theorem 3.2.6 can be used to find a formula for $1/\beta$, as the following example shows.

3.2.7 Example. In $\mathbb{Q}(\sqrt{2})$, if $\beta = 1 + 2\sqrt{2}$ then $\beta^2 = 9 + 4\sqrt{2}$ and it is easy to see that $\beta^2 - 2\beta - 7 = 0$. This can be rewritten as $\beta(\beta - 2) = 7$ so that
$$1/\beta = (\beta - 2)/7 = -\frac{1}{7} + \frac{2}{7}\sqrt{2}.$$
∎

3.2.8 Theorem. [Smallest Field Theorem] *Let $\mathbb{F}$ be a subfield of $\mathbb{C}$ and let $\alpha \in \mathbb{C}$ be algebraic over $\mathbb{F}$. Then $\mathbb{F}(\alpha)$ is the smallest field containing α and all the numbers in $\mathbb{F}$.*

Proof. This is analogous to the proof of Proposition 3.1.5. ∎

The Smallest Field Theorem gives a way of describing $\mathbb{F}(\alpha)$ which is just as useful for many purposes as the definition of $\mathbb{F}(\alpha)$ (Definition 3.2.1). Hence, in solving problems involving $\mathbb{F}(\alpha)$ you should keep both Definition 3.2.1 and Theorem 3.2.8 in mind and then use whichever seems more appropriate for the particular problem.

In some books the rôles of our Smallest Field Theorem and our Definition 3.2.1 are reversed, the statement of the former appearing as a *definition* and the latter as a *theorem*. Our treatment is perhaps more concrete and so hopefully easier for those studying the topic for the first time.

Our approach assumes, however, that α is algebraic over $\mathbb{F}$. If this is not the case, our definition of $\mathbb{F}(\alpha)$ is not applicable. Hence, in such cases, we would use the statement of Theorem 3.2.8 as our definition.

The following example is a typical application of the Smallest Field Theorem.

3.2.9 Example. $\mathbb{F}(\alpha^2) \subseteq \mathbb{F}(\alpha)$ for each subfield $\mathbb{F}$ of $\mathbb{C}$ and each $\alpha \in \mathbb{C}$.

Proof. Let $\mathbb{F}$ be a subfield of $\mathbb{C}$ and let $\alpha \in \mathbb{C}$.
By the Smallest Field Theorem, $\mathbb{F}(\alpha)$ contains α and $\mathbb{F}$.
Hence $\mathbb{F}(\alpha)$ contains $\alpha.\alpha = \alpha^2$ since, being a field, $\mathbb{F}(\alpha)$ is closed under multiplication.
This shows that $\mathbb{F}(\alpha)$ is a field containing α^2 and $\mathbb{F}$.
But $\mathbb{F}(\alpha^2)$ is the smallest such field.
Therefore $\mathbb{F}(\alpha^2) \subseteq \mathbb{F}(\alpha)$, as required. ∎

Extending Fields

Exercises 3.2

1. What does the definition of $\mathbb{F}(\alpha)$ say in each of the following cases?
   (i) $\deg(\alpha, \mathbb{F}) = 1$,
   (ii) $\deg(\alpha, \mathbb{F}) = 2$,
   (iii) $\deg(\alpha, \mathbb{F}) = 3$.

2. (a) What does the definition of $\mathbb{F}(\alpha)$ tell you in the special case where $\mathbb{F} = \mathbb{Q}$ and $\alpha = \sqrt{3}$?
   (b) What do Propositions 3.2.2, 3.2.3 and Theorem 3.2.4 tell you in this special case?
   (c) Verify directly that
     (i) the set $\{1, \sqrt{3}\}$ is linearly independent over $\mathbb{Q}$,
     (ii) $\mathbb{Q}(\sqrt{3})$ contains the product $(2 + \sqrt{3})(3 + \sqrt{3})$ and the quotient $(2 + \sqrt{3})/(3 + \sqrt{3})$.

3. What is the degree of $\sqrt{2}$ over $\mathbb{R}$? Which familiar field is $\mathbb{R}(\sqrt{2})$?

4. What is the degree of i over $\mathbb{R}$? Which familiar field is $\mathbb{R}(i)$?

5. (a) State why $1 + \sqrt{2} \in \mathbb{Q}(\sqrt{2})$ and why $\sqrt{2} \in \mathbb{Q}(1 + \sqrt{2})$.
   (b) Prove that $\mathbb{Q}(1 + \sqrt{2}) = \mathbb{Q}(\sqrt{2})$ by showing that each field is a subset of the other. (You may use the Smallest Field Theorem.)

6. (a) Write down the irreducible polynomial of $\sqrt{3}$ over each of the following fields:
     (i) $\mathbb{Q}$,   (ii) $\mathbb{R}$,   (iii) $\mathbb{Q}(\sqrt{3})$.
   (b) State the degree of $\sqrt{3}$ over each of these fields.

7. Find $\mathrm{irr}(\sqrt{2}, \mathbb{Q}(1 + \sqrt{2}))$ and $\deg(\sqrt{2}, \mathbb{Q}(1 + \sqrt{2}))$. (Justify your answer.)

8. (a) Prove that $\sqrt{2} \notin \mathbb{Q}(\sqrt{3})$.
   (b) Show that $\sqrt{2}$ is algebraic over $\mathbb{Q}(\sqrt{3})$.
   (c) Write down $\mathrm{irr}(\sqrt{2}, \mathbb{Q}(\sqrt{3}))$.
   (d) Justify your answer to (c). (Use (a).)

9. (a) What does your answer to Exercise 8(c) tell you about deg($\sqrt{2}$, $\mathbb{Q}(\sqrt{3})$)?
   (b) Let $\mathbb{F}$ denote $\mathbb{Q}(\sqrt{3})$. Use (a) to write down a basis for the vector space $\mathbb{F}(\sqrt{2})$ over $\mathbb{F}$.

10. Complete the proof of Proposition 3.2.5 by showing that $\mathbb{F}(\alpha)$ is closed under addition.

11. Let $\mathbb{F}$ be a subfield of $\mathbb{C}$ and let $\alpha, \beta \in \mathbb{C}$.
    (a) Prove that if $\mathbb{F}(\beta) \subseteq \mathbb{F}(\alpha)$ then $\beta \in \mathbb{F}(\alpha)$.
    (b) Prove the converse of the result stated in part (a).
    [Hint. Use the Smallest Field Theorem.]
    (c) Deduce that $\mathbb{F}(\alpha) = \mathbb{F}(\beta)$ if and only if $\alpha \in \mathbb{F}(\beta)$ and $\beta \in \mathbb{F}(\alpha)$.

12. Let $\beta = 1 + \sqrt[3]{2}$. Follow the method used in the proof of Theorem 3.2.6 and in Example 3.2.7 to obtain a formula in $\mathbb{Q}(\sqrt[3]{2})$ for $1/\beta$.

13. Let $\mathbb{F}$ be a subfield of $\mathbb{C}$ and let α be a nonzero complex number. Show that if $1/\alpha$ can be expressed in the form
    $$d_0 + d_1\alpha + \ldots + d_k\alpha^k$$
    for $k \geq 1$ and $d_i \in \mathbb{F}$, then α is algebraic over $\mathbb{F}$.
    [Remark. It follows that if $\mathbb{F}(\alpha)$ is defined to be the linear span of all positive powers $1, \alpha, \alpha^2, \ldots$ of α (much as in Definition 3.2.1) then $\mathbb{F}(\alpha)$ is a field precisely when α is algebraic over $\mathbb{F}$.]

14.* Find all the fields $\mathbb{F}$ such that $\mathbb{F}(\sqrt{2}) = \mathbb{Q}(\sqrt{2})$. (Justify your answer.)

15.* Describe the field $\mathbb{Q}(\pi)$ as explicitly as possible.
    [Note that π is not algebraic over $\mathbb{Q}$ and so Definition 3.2.1 is not applicable. Instead, $\mathbb{Q}(\pi)$ is defined as the smallest subfield of $\mathbb{C}$ which contains $\mathbb{Q}$ and π.]

3.3 Iterating the Construction

The starting point for the previous section was a field $\mathbb{F} \subseteq \mathbb{C}$ and a number $\alpha \in \mathbb{C}$ which was algebraic over $\mathbb{F}$. From these ingredients a new field was constructed,
$$\mathbb{F}(\alpha) \subseteq \mathbb{C}.$$

Extending Fields

We can now take the field $F(\alpha)$ and a number $\beta \in \mathbb{C}$ which is algebraic over $F(\alpha)$ as the starting point for a further application of the construction process to get a further new field

$$F(\alpha)(\beta) \subseteq \mathbb{C}.$$

This process can be repeated as often as we like to give a "tower" of fields, each inside the next,

$$\mathbb{Q} \subseteq F \subseteq F(\alpha) \subseteq F(\alpha)(\beta) \subseteq F(\alpha)(\beta)(\gamma) \subseteq \ldots \subseteq \mathbb{C}.$$

3.3.1 Example. $\mathbb{Q}(\sqrt{2})(\sqrt{3})$ is the linear span over $\mathbb{Q}(\sqrt{2})$ of the set of vectors $\{1, \sqrt{3}\}$. This set, furthermore, is a basis for the vector space $\mathbb{Q}(\sqrt{2})(\sqrt{3})$ over $\mathbb{Q}(\sqrt{2})$.

Proof. First note that by the solution to Example 3.1.6,

$$\mathrm{irr}(\sqrt{3}, \mathbb{Q}(\sqrt{2})) = X^2 - 3$$

and hence

$$\deg(\sqrt{3}, \mathbb{Q}(\sqrt{2})) = 2.$$

It now follows from Definition 3.2.1 that

$$\mathbb{Q}(\sqrt{2})(\sqrt{3}) = \{x + \sqrt{3}y : x, y \in \mathbb{Q}(\sqrt{2})\},$$

which is just the linear span of the set of vectors $\{1, \sqrt{3}\}$ over $\mathbb{Q}(\sqrt{2})$. That this set of vectors forms a basis follows from Theorem 3.2.4. ∎

The tower of fields

$$\mathbb{Q} \subseteq \mathbb{Q}(\sqrt{2}) \subseteq \mathbb{Q}(\sqrt{2})(\sqrt{3})$$

invites us to consider $\mathbb{Q}(\sqrt{2})(\sqrt{3})$ as a vector space over $\mathbb{Q}$ also. This leads to the following example.

3.3.2 Example. $\mathbb{Q}(\sqrt{2})(\sqrt{3})$ is the linear span over $\mathbb{Q}$ of the set of vectors $\{1, \sqrt{2}, \sqrt{3}, \sqrt{2}\sqrt{3}\}$.

Proof. By the previous example

$$\begin{aligned}
\mathbb{Q}(\sqrt{2})(\sqrt{3}) &= \{x + \sqrt{3}y : x, y \in \mathbb{Q}(\sqrt{2})\} \\
&= \{(a + b\sqrt{2}) + \sqrt{3}(c + d\sqrt{2}) : a, b, c, d \in \mathbb{Q}\} \\
&= \{a + b\sqrt{2} + c\sqrt{3} + d\sqrt{2}\sqrt{3} : a, b, c, d \in \mathbb{Q}\}
\end{aligned}$$

which expresses it as the required linear span. ∎

One might guess from the above example that the set of vectors $\{1, \sqrt{2}, \sqrt{3}, \sqrt{2}\sqrt{3}\}$ is in fact a basis for the vector space $\mathbb{Q}(\sqrt{2})(\sqrt{3})$ over $\mathbb{Q}$. One of the aims of the next section is to prove a theorem from which this result will follow very easily.

――――――――――――――――― **Exercises 3.3** ―――――――――――――――――

1. Find a spanning set for $\mathbb{Q}(\sqrt{2})(i)$
   (a) as a vector space over $\mathbb{Q}(\sqrt{2})$,
   (b) as a vector space over $\mathbb{Q}$.

2. Simplify $\mathbb{F}(\alpha)(1+\alpha)$, where $\mathbb{F}$ is a subfield of $\mathbb{C}$ and $\alpha \in \mathbb{C}$. (Justify your answer.)

3. Write down three different extension fields of $\mathbb{Q}$ which contain $\sqrt{2}$.

4. If $\mathbb{F}$ is a subfield of $\mathbb{C}$ and $\alpha, \beta \in \mathbb{C}$ then $\mathbb{F}(\alpha, \beta)$ is defined to be the smallest subfield of $\mathbb{C}$ which contains all numbers in $\mathbb{F}$ and also the numbers α and β. Use this definition to show that, if α and β are algebraic over $\mathbb{F}$, then

$$\mathbb{F}(\alpha)(\beta) = \mathbb{F}(\alpha, \beta) = \mathbb{F}(\beta)(\alpha)$$

where, as in this section, $\mathbb{F}(\alpha)(\beta)$ means the smallest field containing all numbers in the field $\mathbb{F}(\alpha)$ and the number β.

3.4 Towers of Fields

The aim of this section is to produce a theorem which is of vital importance in questions involving a tower

$$\mathbb{E} \subseteq \mathbb{F} \subseteq \mathbb{K}$$

consisting of three distinct subfields $\mathbb{E}, \mathbb{F}$ and $\mathbb{K}$ of $\mathbb{C}$. Implicit in this set-up are three different vector spaces and the theorem to be proved will give a precise relationship between their dimensions.

The three vector spaces arising from the tower are as follows. Since $\mathbb{E}$ is a subfield of $\mathbb{F}$, we may take $\mathbb{F}$ as the vectors and $\mathbb{E}$ as the scalars to give the vector space

(i) $\mathbb{F}$ over $\mathbb{E}$.

Likewise there are the vector spaces

Extending Fields

(ii) **K** over **F**, and

(iii) **K** over **E**.

It may be helpful to see all these vector spaces together on a single diagram, as shown below.

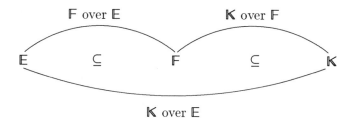

Here **F** plays a schizophrenic rôle: looking back we regard its elements as vectors, but looking forward we take them as scalars.

The theorem on dimensions will follow easily from the following theorem.

3.4.1 Theorem. [Basis for a Tower Theorem] *Consider a tower of subfields of* $\mathbb{C}$,

$$\mathbb{E} \subseteq \mathbb{F} \subseteq \mathbb{K}.$$

If the vector space **F** *over* **E** *has a basis* $\{\alpha_1, \ldots, \alpha_m\}$ *and the vector space* **K** *over* **F** *has a basis* $\{\beta_1, \ldots, \beta_n\}$, *then the set of vectors*

$$\begin{array}{cccc}
\alpha_1\beta_1, & \alpha_1\beta_2, & \ldots, & \alpha_1\beta_n, \\
\alpha_2\beta_1, & \alpha_2\beta_2, & \ldots, & \alpha_2\beta_n, \\
\vdots & \vdots & & \vdots \\
\alpha_m\beta_1, & \alpha_m\beta_2, & \ldots, & \alpha_m\beta_n
\end{array}$$

forms a basis for the vector space **K** *over* **E**.

Proof. The following diagram may help you to keep track of where the vectors come from.

$$\mathbb{E} \underset{\alpha_j \in \mathbb{F}}{\subseteq} \mathbb{F} \underset{\substack{\alpha_j\beta_i \in \mathbb{K} \\ \beta_i \in \mathbb{K}}}{\subseteq} \mathbb{K}$$

Firstly we show that the given set of vectors *spans* the vector space **K** over **E**. To do this, let $k \in \mathbb{K}$.

> Our aim is to prove that k is a linear combination of the $\alpha_j\beta_i$'s with coefficients in $\mathbb{E}$.

Because the β's form a basis for $\mathbb{K}$ over $\mathbb{F}$, there exist n scalars $f_1, \ldots, f_n \in \mathbb{F}$ such that
$$k = \sum_{i=1}^{n} f_i \beta_i. \tag{1}$$
But since the α's form a basis for $\mathbb{F}$ over $\mathbb{E}$, there exist scalars e_{ij} $(1 \leq i \leq n, 1 \leq j \leq m)$ in $\mathbb{E}$ such that
$$f_i = \sum_{j=1}^{m} e_{ij}\alpha_j. \tag{2}$$
Substituting (2) in (1) gives
$$k = \sum_{i=1}^{n} \left(\sum_{j=1}^{m} e_{ij}\alpha_j \right) \beta_i$$
$$= \sum_{i=1}^{n} \sum_{j=1}^{m} e_{ij}(\alpha_j \beta_i).$$
Thus the $\alpha_j\beta_i$'s span the vector space $\mathbb{K}$ over $\mathbb{E}$.

Secondly we show that these vectors are *linearly independent* over $\mathbb{E}$. Suppose e_{ij} $(1 \leq i \leq n, 1 \leq j \leq m)$ are elements of $\mathbb{E}$ such that
$$\sum_{i=1}^{n} \sum_{j=1}^{m} e_{ij}\alpha_j\beta_i = 0. \tag{3}$$

> Our aim is to prove that all the e's are zero.

We can rewrite (3) as
$$\sum_{i=1}^{n} \left(\sum_{j=1}^{m} e_{ij}\alpha_j \right) \beta_i = 0.$$
But in this sum, the coefficients of the β_i's (each coefficient is a sum of the $e_{ij}\alpha_j$'s) are elements of $\mathbb{F}$ and the β's are linearly independent over $\mathbb{F}$. Hence, for $1 \leq i \leq n$,
$$\sum_{j=1}^{m} e_{ij}\alpha_j = 0.$$

Extending Fields

But the α's are linearly independent over E, which means that, for $1 \le j \le m$,
$$e_{ij} = 0.$$
This establishes the required linear independence.

Thus the $\alpha_j \beta_i$'s span the vector space K over E and are linearly independent over E. Hence they form a basis for the vector space K over E. ∎

3.4.2 Example. The vector space $\mathbb{Q}(\sqrt{2})(\sqrt{3})$ over $\mathbb{Q}$ has the set of vectors
$$\{1, \sqrt{2}, \sqrt{3}, \sqrt{2}\sqrt{3}\}$$
as a basis.

Proof. Consider the tower
$$\mathbb{Q} \subseteq \mathbb{Q}(\sqrt{2}) \subseteq \mathbb{Q}(\sqrt{2})(\sqrt{3}).$$
By the Basis for $\mathsf{F}(\alpha)$ Theorem (Theorem 3.2.4) and Example 2.3.3, the vector space $\mathbb{Q}(\sqrt{2})$ over $\mathbb{Q}$ has a basis $\{1, \sqrt{2}\}$ while by the same theorem and Example 3.1.6, the vector space $\mathbb{Q}(\sqrt{2})(\sqrt{3})$ over $\mathbb{Q}(\sqrt{2})$ has a basis $\{1, \sqrt{3}\}$. Hence, by the Basis for a Tower Theorem (Theorem 3.4.1), the vector space $\mathbb{Q}(\sqrt{2})(\sqrt{3})$ over $\mathbb{Q}$ has the basis
$$\{1, \sqrt{2}, \sqrt{3}, \sqrt{2}\sqrt{3}\}. \qquad \blacksquare$$

By counting the numbers of elements in the bases for the vector spaces occurring in Theorem 3.4.1 we can get a relationship between their dimensions. This gives us the following very useful theorem.

3.4.3 Theorem. [Dimension for a Tower Theorem] *Consider a tower of subfields of $\mathbb{C}$,*
$$\mathsf{E} \subseteq \mathsf{F} \subseteq \mathsf{K}.$$
If the vector spaces F over E and K over F have finite dimension then so does K over E and
$$[\mathsf{K} : \mathsf{E}] = [\mathsf{K} : \mathsf{F}][\mathsf{F} : \mathsf{E}]$$
or, in alternative notation,
$$\dim_{\mathsf{E}} \mathsf{K} = \dim_{\mathsf{F}} \mathsf{K} \cdot \dim_{\mathsf{E}} \mathsf{F}.$$

Proof. There are altogether mn of the $\alpha\beta$'s in Theorem 3.4.1 and these form a basis for $\mathbb{K}$ over $\mathbb{E}$. This vector space therefore has dimension mn. ∎

Since the dimension of a vector space must be an integer, Theorem 3.4.3 has the following corollary.

3.4.4 Corollary. *Under the conditions of Theorem 3.4.3, $[\mathbb{K}:\mathbb{F}]$ is a factor of $[\mathbb{K}:\mathbb{E}]$ and $[\mathbb{F}:\mathbb{E}]$ is a factor of $[\mathbb{K}:\mathbb{E}]$.* ∎

As an application of this corollary, observe that there cannot be any field $\mathbb{F}$ such that

$$\mathbb{Q} \subseteq \mathbb{F} \subseteq \mathbb{Q}(\sqrt{2})(\sqrt{3})$$

with $[\mathbb{F}:\mathbb{Q}] = 3$ since, by Example 3.4.2,

$$[\mathbb{Q}(\sqrt{2})(\sqrt{3}):\mathbb{Q}] = 4$$

and 3 is not a factor of 4.

──────────── **Exercises 3.4** ────────────

1. How many vector spaces can be constructed from a tower of four distinct fields
   $$\mathbb{E} \subseteq \mathbb{F} \subseteq \mathbb{G} \subseteq \mathbb{K}$$
   by using larger fields as vector spaces over smaller fields?

2. Consider a tower of subfields of $\mathbb{C}$,
   $$\mathbb{E} \subseteq \mathbb{F} \subseteq \mathbb{K}.$$
   (a) What is $[\mathbb{K}:\mathbb{E}]$ if $[\mathbb{K}:\mathbb{F}] = 2$ and $[\mathbb{F}:\mathbb{E}] = 6$?
   (b) Is it possible that $[\mathbb{K}:\mathbb{E}] = 3$ and $[\mathbb{F}:\mathbb{E}] = 2$? (Justify your answer.)

3. Give an example of a tower of three distinct subfields of $\mathbb{C}$,
   $$\mathbb{E} \subseteq \mathbb{F} \subseteq \mathbb{K},$$
   such that $[\mathbb{K}:\mathbb{E}] = 4$.

4. By using the tower

$$\mathbb{Q} \subseteq \mathbb{Q}(\sqrt{3}) \subseteq \mathbb{Q}(\sqrt{3})(\sqrt{2})$$

write down a basis for the vector space

$$\mathbb{Q}(\sqrt{3})(\sqrt{2}) \text{ over } \mathbb{Q}.$$

> Exercises 5 to 10 will refer to the tower
>
> $$\mathbb{Q} \subseteq \mathbb{Q}(\sqrt{6}) \subseteq \mathbb{Q}(\sqrt{2}+\sqrt{3}).$$

5. (a) Is the set $\{1, \sqrt{2}, \sqrt{3}, \sqrt{6}\}$ linearly independent over $\mathbb{Q}$? Why? [Hint. Use your answer to Exercise 4.]
   (b) Does $\sqrt{2}+\sqrt{3} \in \mathbb{Q}(\sqrt{6})$? Why?
   (c) Does $\sqrt{6} \in \mathbb{Q}(\sqrt{2}+\sqrt{3})$? Why? [Hint. Calculate $(\sqrt{2}+\sqrt{3})^2$.]

6. Which of your answers to Exercise 5 would you use
   (i) to validate the above tower?
   (ii) to show that $\mathbb{Q}(\sqrt{6})(\sqrt{2}+\sqrt{3}) = \mathbb{Q}(\sqrt{2}+\sqrt{3})$?

7. (a) Show that the number $\sqrt{2}+\sqrt{3}$ is algebraic over $\mathbb{Q}(\sqrt{6})$.
   (b) Write down irr$(\sqrt{2}+\sqrt{3}, \mathbb{Q}(\sqrt{6}))$ and state which fact from Exercise 5 you would need to use to justify your answer.
   (c) Deduce the value of $[\mathbb{Q}(\sqrt{6})(\sqrt{2}+\sqrt{3}) : \mathbb{Q}(\sqrt{6})]$.
   (d) Deduce the value of $[\mathbb{Q}(\sqrt{2}+\sqrt{3}) : \mathbb{Q}(\sqrt{6})]$.

8. Use your answer to Exercise 7(d) and the Dimension for a Tower Theorem to deduce
   (i) $[\mathbb{Q}(\sqrt{2}+\sqrt{3}) : \mathbb{Q}]$,
   (ii) a basis for the vector space $\mathbb{Q}(\sqrt{2}+\sqrt{3})$ over $\mathbb{Q}$.

9. Let $\beta \in \mathbb{Q}(\sqrt{2}+\sqrt{3})$.
   (a) Use the fact that any $m+1$ vectors in a vector space of dimension m are linearly dependent, and your answer to Exercise 8(i), to show that $\{1, \beta, \beta^2, \beta^3, \beta^4\}$ is linearly dependent over $\mathbb{Q}$.

(b) Deduce from (a) that β is algebraic over $\mathbb{Q}$.

(c) What can you say about $\deg(\beta, \mathbb{Q})$?

10. (a) Use your answer to Exercise 8(ii) and the dimension part of the Basis for $F(\alpha)$ Theorem (Theorem 3.2.4) to deduce the value of $\deg(\sqrt{2}+\sqrt{3}, \mathbb{Q})$. [Hint. Remember that any two bases of a vector space contain the same number of vectors.]

(b) Use (a) and the Basis for $F(\alpha)$ Theorem to write down a basis for $\mathbb{Q}(\sqrt{2}+\sqrt{3})$ over $\mathbb{Q}$. Is the fact that this is different from the one in Exercise 8(ii) a problem?

Exercises 11 to 15 will refer to the tower

$$\mathbb{Q} \subseteq \mathbb{Q}(\sqrt{2}) \subseteq \mathbb{Q}(\sqrt[4]{2}).$$

---

11. (a) Show that $\sqrt[4]{2}$ is algebraic over $\mathbb{Q}$. Is it also algebraic over $\mathbb{Q}(\sqrt{2})$?

(b) Does $\sqrt[4]{2} \in \mathbb{Q}(\sqrt{2})$? (Justify your answer.)

(c) Does $\sqrt{2} \in \mathbb{Q}(\sqrt[4]{2})$? (Justify your answer.)

12. Which of your answers to Exercise 11 would you use

(ii) to validate the above tower?

(ii) to show that $\mathbb{Q}(\sqrt{2})(\sqrt[4]{2}) = \mathbb{Q}(\sqrt[4]{2})$?

13. (a) Write down the polynomial $\mathrm{irr}(\sqrt[4]{2}, \mathbb{Q}(\sqrt{2}))$ and state which fact from Exercise 11 you would use to justify your answer.

(b) Deduce the value of $[\mathbb{Q}(\sqrt{2})(\sqrt[4]{2}) : \mathbb{Q}(\sqrt{2})]$.

(c) Deduce the value of $[\mathbb{Q}(\sqrt[4]{2}) : \mathbb{Q}(\sqrt{2})]$.

14. Use your answer to Exercise 13(c) and the Dimension for a Tower Theorem to deduce

(i) the value of $[\mathbb{Q}(\sqrt[4]{2}) : \mathbb{Q}]$,

(ii) a basis for the vector space $\mathbb{Q}(\sqrt[4]{2})$ over $\mathbb{Q}$,

(iii) the polynomial $\mathrm{irr}(\sqrt[4]{2}, \mathbb{Q})$.

Extending Fields 59

15. Let F be a subfield of $\mathbb{C}$ and let α be a real number which is algebraic over F of degree n.
    (a) What is $[\mathsf{F}(\alpha):\mathsf{F}]$?
    (b) What is the maximum number of vectors in a linearly independent subset of the vector space $\mathsf{F}(\alpha)$ over F?
    (c) Assume now that $\beta \in \mathsf{F}(\alpha)$.
        (i) Show that β is algebraic over F.
        (ii) Use a suitable tower to prove that the degree of β over F is a factor of that of α over F.
    (d) Use your answer to (c) to show that every number in $\mathbb{Q}(\sqrt[4]{2})$ is algebraic over $\mathbb{Q}$ and has degree 1, 2 or 4 over $\mathbb{Q}$.

Additional Reading for Chapter 3

Our treatment of $\mathsf{F}(\alpha)$ uses relatively little algebraic machinery. Other treatments rely on the algorithm for finding the greatest common divisor of two polynomials (see, for example, Section 110 of [AC], Section 2.2 of [CH], or Section 10.1 of [LS]) or they rely on results about maximal ideals in polynomial rings (see, for example, Section 35 of [JF], or Chapter 11 of [WG]). The proof of the Basis for a Tower Theorem is quite standard and can also be found in Section 97 of [AC], Section 38 of [JF], and Section 2.2 of [CH].

[AC] A. Clark, *Elements of Abstract Algebra*, Wadsworth, Belmont, California, 1971.

[JF] J.B. Fraleigh, *A First Course in Abstract Algebra*, 3rd edition, Addison-Wesley, Reading, Massachusetts, 1982.

[WG] W.J. Gilbert, *Modern Algebra with Applications*, Wiley, New York, 1976.

[CH] C.R. Hadlock, *Field Theory and its Classical Problems*, Carus Mathematical Monographs, No. 19, Mathematical Association of America, 1978.

[LS] L.W. Shapiro, *Introduction to Abstract Algebra*, McGraw-Hill, New York, 1975.

CHAPTER 4

Irreducible Polynomials

In our earlier definition of the irreducible polynomial of a number, the word "irreducible" was intended to convey the idea that the *degree* of the polynomial "could not be reduced further". In this chapter it will be shown that this polynomial is also "irreducible" in the sense that it "cannot be *factorized* further". This will lead to a practical technique for finding the irreducible polynomial of a number.

With the earlier definition we were able to show, for example, that the irreducible polynomial of $\sqrt{2}$ over $\mathbb{Q}$ is $X^2 - 2$. The techniques to be developed in this chapter will enable us to move beyond quadratics and to show, for example, that

$$\mathrm{irr}(\sqrt[3]{2}, \mathbb{Q}) = X^3 - 2.$$

This will feature later in our proof of the impossibility of doubling the cube.

4.1 Irreducible Polynomials

In this section the concept of an "irreducible polynomial" will be defined. You have already met this phrase in the combination "the irreducible polynomial of a number α". Whereas the earlier concept was defined in terms of "least degree", we shall now turn to the idea of "cannot be further factorized". The precise relationship between an

"irreducible polynomial"

and

"the irreducible polynomial of a number α"

will be explained later, in Section 4.3. For the moment, however, the main thing is that you do not confuse the two concepts.

4.1.1 Definition. Let $\mathbb{F}$ be a field. A polynomial $f(X) \in \mathbb{F}[X]$ is said to be *reducible over* $\mathbb{F}$ if there are polynomials $g(X)$ and $h(X)$ in $\mathbb{F}[X]$ such that

(i) each has degree less than that of $f(X)$, and

(ii) $f(X) = g(X)h(X)$. ∎

Thus a polynomial is reducible if it can be factorized in a suitable way.

4.1.2 Example. The polynomial $X^2 - 2$ is reducible over $\mathbb{R}$.

Proof. Observe that $X^2 - 2 = (X - \sqrt{2})(X + \sqrt{2})$, where each of the factors belongs to $\mathbb{R}[X]$ and has degree less than that of $X^2 - 2$. ∎

4.1.3 Example. The polynomial $X^2 - 2$ is not reducible over $\mathbb{Q}$.

Proof. To see this, suppose, on the contrary, that $X^2 - 2$ is reducible over $\mathbb{Q}$. This means that

$$X^2 - 2 = (aX + b)(cX + d)$$

for some $a, b, c, d \in \mathbb{Q}$. Clearly neither a nor c can be zero. Evaluating both sides at $\sqrt{2}$ gives

$$0 = (a\sqrt{2} + b)(c\sqrt{2} + d).$$

Hence $\sqrt{2} = -b/a$ or $-d/c$, which gives the contradiction that $\sqrt{2}$ is rational. ∎

Irreducible Polynomials

4.1.4 Example. Constant polynomials and polynomials of degree 1 are never reducible. ∎

Linguistically speaking, the word "irreducible" should simply mean not reducible, because the prefix ir means not.
(For example,

ir̲replaceable means "not replaceable",

ir̲responsible means "not responsible",

ir̲religious means " not religious".)

This is nearly true in the following definition, but there is a slight twist in that constant polynomials are excluded.

4.1.5 Definition. Let $\mathbb{F}$ be a field. A polynomial $f(X)$ in $\mathbb{F}[X]$ is said to be *irreducible over* $\mathbb{F}$ if it is not reducible over $\mathbb{F}$ and it is not a constant. ∎

Thus a polynomial is irreducible if it cannot be factorized suitably. The reason for excluding constant polynomials in this definition is that, without this exclusion, certain theorems which appear later would need to be stated in a more cumbersome way.

In terms of "irreducible", the examples given earlier show that

$X^2 - 2$ *is not irreducible over* $\mathbb{R}$,

$X^2 - 2$ *is irreducible over* $\mathbb{Q}$.

Note that if a polynomial is irreducible over a field, then it is irreducible over all subfields of that field. More precisely, if $\mathbb{E}$ is a subfield of $\mathbb{F}$ and $f(X) \in \mathbb{E}[X]$ is irreducible over $\mathbb{F}$, then $f(X)$ is irreducible over $\mathbb{E}$.

Notice also that constant polynomials are neither reducible nor irreducible; but all other polynomials are either reducible or irreducible over a given field.

──────────── **Exercises 4.1** ────────────

1. In each case state (giving reasons) whether the polynomial is reducible, irreducible or neither over the given field.
   (a) the polynomial $X^2 + 3X + 2$ (i) over $\mathbb{R}$, (ii) over $\mathbb{Q}$;

(b) the polynomial $X^2 + 1$
   (i) over $\mathbb{Q}$,   (ii) over $\mathbb{C}$,   (iii) over $\mathbb{Z}_2$,   (iv) over $\mathbb{Z}_3$;
(c) the polynomial 10 over $\mathbb{Q}$;
(d) the polynomial $X^2 - 2$ (i) over $\mathbb{Q}$,   (ii) over $\mathbb{Q}(\sqrt{2})$.

2. Does the fact that $3X^2 + 3 = 3(X^2 + 1)$ in $\mathbb{Q}[X]$ mean that $3X^2 + 3$ is reducible over $\mathbb{Q}$?

3. (a) Is $X^3 - 1$ irreducible over $\mathbb{Q}$?
   (b) Is $X^3 + 1$ irreducible over $\mathbb{Q}$?
   (c) Is $X^{666} - 1$ irreducible over $\mathbb{Q}$?

4. Let $f(X) \in \mathbb{F}[X]$ be irreducible over $\mathbb{F}$ and let $g(X), h(X)$ in $\mathbb{F}[X]$ be such that $f(X) = g(X) h(X)$. Show that either $g(X)$ or $h(X)$ is a constant polynomial.

5. Assume that $\alpha \in \mathbb{C}$ is algebraic over a field $\mathbb{F} \subseteq \mathbb{C}$ and that $f(X) = \mathrm{irr}(\alpha, \mathbb{F})$. Show that $f(X)$ is not reducible over $\mathbb{F}$.

   [Hint. Use the fact that $f(X)$ is of least possible degree amongst the polynomials in $\mathbb{F}[X]$ with α as a zero, and the fact that if $pq = 0$ for p, q in $\mathbb{C}$ then either $p = 0$ or $q = 0$.]

4.2 Reducible Polynomials and Zeros

In this section the relationship between a polynomial having a zero and being reducible will be explored. This relationship will form the basis of our technique for proving the irreducibility of certain polynomials.

As a first step in this direction, the relationship between having a zero and having a factor of degree 1 will be explored.

4.2.1 Definition. Let $\mathbb{F}$ be a field. A polynomial $f(X) \in \mathbb{F}[X]$ is said to have a *factor of degree 1 in* $\mathbb{F}[X]$ if

$$f(X) = (aX + b) g(X)$$

where $a, b \in \mathbb{F}$ with $a \neq 0$ and where $g(X) \in \mathbb{F}[X]$.   ∎

Irreducible Polynomials

4.2.2 Theorem. [Factor Theorem] *Let $\mathbb{F}$ be a field. A polynomial $f(X) \in \mathbb{F}[X]$ has a factor of degree 1 in $\mathbb{F}[X]$ if and only if $f(X)$ has a zero in $\mathbb{F}$.*

Proof. Assume firstly that $f(X)$ has a factor of degree 1. With notation as in Definition 4.2.1, it follows that

$$-b/a \in \mathbb{F}$$

is a zero of $f(X)$ since

$$f(-b/a) = \bigl(a(-b/a) + b\bigr)g(-b/a)$$
$$= 0.$$

Conversely, assume that $\alpha \in \mathbb{F}$ is a zero of $f(X)$.

> We aim to show that $X - \alpha$ is a factor of $f(X)$.

If we divide $f(X)$ by $X - \alpha$ then (by the Division Theorem, Theorem 1.3.2) there exist $q(X), r(X)$ in $\mathbb{F}[X]$ with

$$f(X) = (X - \alpha)\, q(X) + r(X) \tag{1}$$

and

$$\text{either } r(X) = 0 \text{ or } \deg r(X) < \deg(X - \alpha) = 1.$$

Since $r(X) = 0$ or its degree is less than 1, $r(X)$ must be a constant polynomial $c \in \mathbb{F} \subseteq \mathbb{F}[X]$, which means we can rewrite (1) as

$$f(X) = (X - \alpha)\, q(X) + c. \tag{2}$$

If we substitute $X = \alpha$ into this and use the fact that $f(\alpha) = 0$ (why?), we have

$$0 = f(\alpha) = (\alpha - \alpha)q(\alpha) + c = 0 + c = c.$$

Thus, from (2), $f(X) = (X - \alpha)q(X)$ which means $f(X)$ has the factor $X - \alpha$ of degree 1 in $\mathbb{F}[X]$. ∎

The most useful theorem for dealing with polynomials of degree 2 or 3 is the following one.

4.2.3   Theorem. [Small Degree Irreducibility Theorem]   Let F be any field. Let $f(X)$ in $\mathsf{F}[X]$ have degree 2 or 3. Then $f(X)$ is reducible over F if and only if $f(X)$ has a zero in F.

Proof.   Let $f(X) \in \mathsf{F}[X]$ have degree 2 or 3.

Assume firstly that $f(X)$ is reducible over F, so that

$$f(X) = g(X) h(X)$$

for some nonconstant polynomials $g(X), h(X) \in \mathsf{F}[X]$. Since the degrees of $g(X)$ and $h(X)$ must add up to 2 or 3, one (or both) of these degrees must be 1. Hence, by Theorem 4.2.2, one of them must have a zero in F; hence $f(X)$ must also have a zero in F.

To prove the converse is easy: assume $f(X)$ has a zero in F and then apply Theorem 4.2.2. ∎

4.2.4   Example. The polynomial $2X^3 - 5$ is irreducible over $\mathbb{Q}$.

Proof.   By the Rational Roots Test (Theorem 1.4.1) the only possible zeros in $\mathbb{Q}$ of the polynomial are

$$\pm 1, \pm \frac{1}{2}, \pm \frac{5}{2}, \pm 5,$$

none of which works (that is, none gives zero when substituted into $2X^3 - 5$).

Thus the polynomial has no zeros in $\mathbb{Q}$. Since its degree is 3, the Small Degree Irreducibility Theorem is applicable and shows that $f(X)$ is irreducible over $\mathbb{Q}$. ∎

It is important to observe that the Small Degree Irreducibility Theorem would not be true if we removed the restriction about "degree 2 or 3". For example, the quartic

$$(X^2 + 1)(X^2 + 4)$$

is reducible over $\mathbb{Q}$ but has no zero in $\mathbb{Q}$.

Fortunately we do not need to worry about the irreducibility of quartics (or higher degree polynomials) in order to prove the impossibility of the three famous geometric constructions.

Exercises 4.2

1. Use the Rational Roots Test and either the Factor Theorem or the Small Degree Irreducibility Theorem to decide which of the polynomials listed below is irreducible over the stated field:
   (a) $X^3 - 2X + 1$ over $\mathbb{Q}$,
   (b) $X^3 + 2X + 1$ over $\mathbb{Q}$,
   (c) $X^3 - 5$ over $\mathbb{Q}$.

2. In which of the following cases can the technique of Exercise 1 be used to decide if the polynomial is irreducible over the given field?
   (a) $X^4 + 2X^3 + 2X^2 + 2X + 1$ over $\mathbb{Q}$,
   (b) $X^4 + 2X^3 - 2X^2 + 2X + 1$ over $\mathbb{Q}$,
   (c) $X^3 - \sqrt{2}X + 1$ over $\mathbb{Q}$,
   (d) $X^2 - 2$ over $\mathbb{Q}$,
   (e) $X^2 - 2$ over $\mathbb{Q}(\sqrt{2})$.

3. In each case find a zero in $\mathbb{C}$ of the given polynomial and then use the method of proof of Theorem 4.2.2 to write down a factor of degree 1 in $\mathbb{C}[X]$ of the given polynomial.
   (a) $10X^2 + 10$,
   (b) $X^2 - 4X + 5$.

4. In each case give an example of a field $\mathbb{F}$ and a polynomial $f(X)$ in $\mathbb{F}[X]$ such that $f(X)$ has no zeros in $\mathbb{F}$ and yet $f(X)$ is not irreducible over $\mathbb{F}$.
   (a) $f(X)$ is constant.
   (b) $f(X)$ is not constant.
   Does this contradict the Small Degree Irreducibility Theorem?

5. In each case decide whether the statement is true or false, where $\mathbb{F}$ is a field. (Justify your answers.)
   (a) If $f(X) \in \mathbb{F}[X]$ is reducible over $\mathbb{F}$ then $f(X)$ has a zero in $\mathbb{F}$.
   (b) If $f(X) \in \mathbb{F}[X]$ has a zero in $\mathbb{F}$ then $f(X)$ is reducible over $\mathbb{F}$.
   (c) If $f(X) \in \mathbb{F}[X]$ has degree 4 and has a zero in $\mathbb{F}$, then $f(X)$ is reducible over $\mathbb{F}$.
   (d) If $f(X) \in \mathbb{F}[X]$ has degree 2 or more and has a zero in $\mathbb{F}$, then $f(X)$ is reducible over $\mathbb{F}$.

6. In each case decide whether the statement is true or false, where F is a field. (Justify your answers.)
    (a) If $f(X) \in F[X]$ has degree 6 and is irreducible over F, then $f(X)$ does not have a zero in F.
    (b) If $f(X) \in F[X]$ is irreducible over F then $f(X)$ does not have a zero in F.
    (c) If $f(X) \in F[X]$ has degree 2 or more and is irreducible over F, then $f(X)$ does not have a zero in F.

4.3 Irreducibility and irr(α, F)

The purpose of this section is to state and prove a theorem which links the idea of "irreducible polynomial" (as in Definition 4.1.5) to the idea of "irreducible polynomial of a number" (as in Definition 2.3.2).

4.3.1 Theorem. [Monic Irreducible Zero Theorem] *Let $F \subseteq C$ be a field and let $\alpha \in C$ be algebraic over F. The following conditions on a polynomial $f(X) \in F[X]$ are equivalent:*
  (i) $f(X) = \mathrm{irr}(\alpha, F)$,
  (ii) $f(\alpha) = 0$ and $f(X)$ is monic and irreducible over F.

Before proving this theorem, we give some examples to show how the theorem can be used to check that a guess for $\mathrm{irr}(\alpha, F)$ is indeed the correct one.

4.3.2 Example. $\mathrm{irr}(\sqrt{2}, Q) = X^2 - 2$.

Proof. Clearly $X^2 - 2$ has $\sqrt{2}$ as a zero and it is monic.

By the Rational Roots Test, the only possible zeros in Q are $\pm 1, \pm 2$, none of which works. Because this polynomial has degree 2, it follows from the Small Degree Irreducibility Theorem that it is irreducible over Q. The Monic Irreducible Zero Theorem now gives the required result. ∎

Actually, this example was done earlier (as Example 2.3.3). The advantage of the Monic Irreducible Zero Theorem is that it lets us handle cubics as well.

Irreducible Polynomials

4.3.3 Example. irr($\sqrt[3]{2}, \mathbb{Q}$) = $X^3 - 2$.

Proof. Clearly $X^3 - 2$ has $\sqrt[3]{2}$ as a zero and it is monic.

By the Rational Roots Test, the only possible zeros in $\mathbb{Q}$ are $\pm 1, \pm 2$, none of which works. Because this polynomial has degree 3, it follows from the Small Degree Irreducibility Theorem that it is irreducible over $\mathbb{Q}$. The Monic Irreducible Zero Theorem now gives the desired result. ∎

Note that the same argument could not be used for quartics because of the requirement on the degree of the polynomial contained in the Small Degree Irreducibility Theorem. Note also that since we are appealing to the Rational Roots Test, the above technique for finding irr($\alpha, \mathbb{F}$) works only in the case $\mathbb{F} = \mathbb{Q}$.

We now complete this section by proving the theorem.

Proof of Theorem 4.3.1. To facilitate the proof, the following notation is introduced:

$$\mathcal{P} = \{p(X) \in \mathbb{F}[X] : p(\alpha) = 0 \text{ and } p(X) \text{ is monic}\}.$$

We first show that **(i) implies (ii)**.

Assume that (i) is true. By Definition 2.3.2 this means that

$$f(X) \text{ is the polynomial of least degree in } \mathcal{P}.$$

Thus the first two statements in (ii) of the theorem hold and it remains to prove that $f(X)$ is irreducible over $\mathbb{F}$.

Suppose $f(X)$ were not irreducible over $\mathbb{F}$. Since it is not constant, it must be reducible over $\mathbb{F}$ and so there are polynomials $g(X)$ and $h(X)$ in $\mathbb{F}[X]$ such that

$$f(X) = g(X) h(X)$$

where the degrees of $g(X)$ and $h(X)$ are both less than that of $f(X)$. We may assume that both of these polynomials are monic. Evaluation of both sides at α gives, furthermore,

$$0 = g(\alpha) h(\alpha)$$

and so, $\mathbb{C}$ being a field, this implies $g(\alpha) = 0$ or $h(\alpha) = 0$. Thus either

$$g(X) \in \mathcal{P} \quad \text{or} \quad h(X) \in \mathcal{P}.$$

But this is a contradiction since $f(X)$ had the least degree in $\mathcal{P}$. Hence $f(X)$ is irreducible over F. So (i) does imply (ii).

Next we show that **(ii) implies (i)**.

Assume that (ii) is true. Thus $f(X) \in \mathcal{P}$ and so

$$\text{degree } f(X) \geq \text{degree irr}(\alpha, \mathsf{F}).$$

By the Division Theorem (Theorem 1.3.2),

$$f(X) = \text{irr}(\alpha, \mathsf{F}) \, q(X) + r(X), \tag{1}$$

where $q(X)$ and $r(X) \in \mathsf{F}(X)$ with $r(X) = 0$ or

$$\text{degree } r(X) < \text{degree irr}(\alpha, \mathsf{F}).$$

Evaluation of the equation (1) at α gives

$$0 = 0.q(\alpha) + r(\alpha)$$

so that $r(\alpha) = 0$. If $r(X) \neq 0$ we can divide $r(X)$ by its leading coefficient to get a monic polynomial in $\mathcal{P}$ whose degree is less than that of $\text{irr}(\alpha, \mathsf{F})$, which is a contradiction. Hence $r(X) = 0$ and so by the equation (1),

$$f(X) = \text{irr}(\alpha, \mathsf{F}) \, q(X).$$

Thus $q(X)$ must be a constant, since otherwise $f(X)$ would not be irreducible over F. Since both $f(X)$ and $\text{irr}(\alpha, \mathsf{F})$ are monic, this means $q(X)$ is 1. ∎

Exercises 4.3
---

1. Find $\text{irr}(\sqrt{5}+1, \mathbb{Q})$ and $\text{irr}(\sqrt[3]{3}-2, \mathbb{Q})$ and then use the method of Examples 4.3.2 and 4.3.3 to justify your answer.

2. (a) Guess $\text{irr}(\sqrt[3]{7}, \mathbb{Q})$ and $\deg(\sqrt[3]{7}, \mathbb{Q})$.
   (b) Now use the definition of $\mathbb{Q}(\sqrt[3]{7})$ to get an explicit description of this set.
   (c) Give a basis for the vector space $\mathbb{Q}(\sqrt[3]{7})$ over $\mathbb{Q}$.
   (d) Give a careful proof, with full justification for each step, for your answer to part (a).

3. Suppose that $\sqrt{7} \in \mathbb{Q}(\sqrt[3]{7})$.
   (a) Deduce that $\mathbb{Q}(\sqrt{7}) \subseteq \mathbb{Q}(\sqrt[3]{7})$.
   (b) Thus we have a tower $\mathbb{Q} \subseteq \mathbb{Q}(\sqrt{7}) \subseteq \mathbb{Q}(\sqrt[3]{7})$.
   Derive a contradiction from this tower.
   (c) What can you now conclude?

4. Note the obvious tower
$$\mathbb{Q} \subseteq \mathbb{Q}(\sqrt[3]{7}) \subseteq \mathbb{Q}(\sqrt[3]{7})(\sqrt{7}).$$
   (a) Guess $\mathrm{irr}(\sqrt{7}, \mathbb{Q}(\sqrt[3]{7}))$ and $\deg(\sqrt{7}, \mathbb{Q}(\sqrt[3]{7}))$.
   (b) Prove your answer to part (a) using the result of Exercise 3.
   (c) Hence give a basis for the two vector spaces
$$\mathbb{Q}(\sqrt[3]{7})(\sqrt{7}) \text{ over } \mathbb{Q}(\sqrt[3]{7}),$$
$$\mathbb{Q}(\sqrt[3]{7})(\sqrt{7}) \text{ over } \mathbb{Q},$$
   and state their dimensions.

5. In each of the following cases show that the number $\sin\theta$ is algebraic over $\mathbb{Q}$ and find its irreducible polynomial and its degree over this field.
   (a) $\theta = \pi/6$,
   (b) $\theta = \pi/3$,
   (c) $\theta = \pi/18$. [Hint. Use the identity $\sin(3\theta) = 3\sin\theta - 4\sin^3\theta$.]

6. Suppose that $\mathbb{E}$ is an extension field of $\mathbb{Q}$, and $\alpha \in \mathbb{E}$ is such that
$$2\alpha^3 + 3\alpha^2 + 4\alpha + 3 = 0.$$
   (a) Write down a monic polynomial $f(X)$ in $\mathbb{Q}[X]$ such that α is a zero of $f(X)$.
   (b) Can we find $\mathrm{irr}(\alpha, \mathbb{Q})$ from the information given? Why?

4.4 Finite-dimensional Extensions

Let $\mathbb{K}$ be an extension field of $\mathbb{F}$ such that the vector space $\mathbb{K}$ over $\mathbb{F}$ is finite-dimensional. Must numbers in $\mathbb{K}$ be algebraic over $\mathbb{F}$ and, if so, what can we say about $\deg(\alpha, \mathbb{F})$ for α in $\mathbb{K}$? The following theorem answers these questions.

4.4.1 Theorem. *Let $\mathbb{F}$ be a subfield of a field $\mathbb{K}$ with $[\mathbb{K}:\mathbb{F}] = n$. Then every number α in $\mathbb{K}$ is algebraic over $\mathbb{F}$ and $\deg(\alpha, \mathbb{F}) \leq n$.*

Proof. Let $\alpha \in \mathbb{K}$. Because $[\mathbb{K}:\mathbb{F}] = n$, every set of $(n+1)$ numbers in $\mathbb{K}$ must be linearly dependent over $\mathbb{F}$. Now

$$\{1, \alpha, \ldots, \alpha^n\}$$

is such a set and thus there exist $c_0, c_1, \ldots c_n \in \mathbb{F}$, not all zero, such that

$$c_0 1 + c_1 \alpha + \ldots + c_n \alpha^n = 0.$$

If we let

$$p(X) = c_0 + c_1 X + \ldots + c_n X^n$$

then $p(X)$ is a nonzero polynomial in $\mathbb{F}[X]$ which has α as a zero. Thus (by definition), α is algebraic over $\mathbb{F}$. It follows easily (using Proposition 2.2.2) that $\deg(\alpha, \mathbb{F}) \leq n$ since $\deg p(X) \leq n$. ∎

4.4.2 Corollary. *Let $\mathbb{F}$ be a subfield of $\mathbb{C}$ and let α be a complex number which is algebraic over $\mathbb{F}$ of degree n. Every number in $\mathbb{F}(\alpha)$ is then algebraic over $\mathbb{F}$ and has degree $\leq n$.* ∎

4.4.3 Theorem. *Let $\mathbb{F}$ be a subfield of a field $\mathbb{E}$. The set of numbers in $\mathbb{E}$ which are algebraic over $\mathbb{F}$ is a subfield of $\mathbb{E}$.*

Proof. Let $\alpha, \beta \in \mathbb{E}$ be algebraic over $\mathbb{F}$. We must show that $\alpha + \beta$, $\alpha - \beta$, $\alpha\beta$ and (provided $\beta \neq 0$) α/β are all algebraic over $\mathbb{F}$. We consider the tower

$$\mathbb{F} \subseteq \mathbb{F}(\alpha) \subseteq \mathbb{F}(\alpha)(\beta).$$

Since α is algebraic over $\mathbb{F}$, $\mathbb{F}(\alpha)$ is a finite-dimensional extension of $\mathbb{F}$. Since β is algebraic over $\mathbb{F}$ it is also algebraic over $\mathbb{F}(\alpha)$ and hence $\mathbb{F}(\alpha)(\beta)$ is a finite-dimensional extension of $\mathbb{F}(\alpha)$. Thus, by the Dimension for a Tower Theorem (Theorem 3.4.3), we see that $\mathbb{F}(\alpha)(\beta)$ is a finite-dimensional extension of $\mathbb{F}$ and so, by Theorem 4.4.1, every element of $\mathbb{F}(\alpha)(\beta)$ is algebraic over $\mathbb{F}$. This completes the proof since $\alpha + \beta$, $\alpha - \beta$, $\alpha\beta$ and (provided $\beta \neq 0$) α/β are all in $\mathbb{F}(\alpha)(\beta)$. ∎

Recall that algebraic numbers are complex numbers which are algebraic over $\mathbb{Q}$.

4.4.4 Corollary. *The algebraic numbers are a subfield of $\mathbb{C}$.* ∎

Irreducible Polynomials

─────────────── **Exercises 4.4** ───────────────

1. Use Theorem 4.4.1 to show that every complex number z is algebraic over $\mathbb{R}$. What can you say about $\deg(z, \mathbb{R})$?

2. (a) Show that, with the hypotheses of Theorem 4.4.1, $\deg(\alpha, \mathsf{F})$ is a factor of n. (Consider the tower $\mathsf{F} \subseteq \mathsf{F}(\alpha) \subseteq \mathsf{K}$.)
   (b) If $\alpha \in \mathbb{Q}(\sqrt[4]{2})$, what can you say about $\deg(\alpha, \mathbb{Q})$?

3. Explain why the number $17\sqrt[4]{2} - 3\sqrt{2}$ is in the field $\mathbb{Q}(\sqrt[4]{2})$ and then use Theorem 4.4.1 to prove that this number is algebraic over the field $\mathbb{Q}$.

4. (a) Explain why $\mathbb{Q}(i)(\sqrt{2})$ is a finite-dimensional extension of $\mathbb{Q}$.
   (b) Use Theorem 4.4.1 to prove that $(3i + 2\sqrt{2})/(1 - 4\sqrt{2}i)$ is algebraic over $\mathbb{Q}$.

5.* Let K be an extension field of $\mathbb{R}$ such that every element α in K is algebraic and of degree at most 2 over $\mathbb{R}$. Prove that $\mathsf{K} = \mathbb{R}$ or $\mathbb{C}$. [Hint. Assume $\deg(\alpha, \mathbb{R}) = 2$. By completing the square in $\mathrm{irr}(\alpha, \mathbb{R})$, show that there are real numbers b and c such that $\beta = (\alpha - b)/c$ satisfies $\beta^2 = -1$.]

6.* Suppose that there exists an extension field K of $\mathbb{R}$ such that $[\mathsf{K} : \mathbb{R}] = 3$.
   (a) Prove that there exists an element α in K which is algebraic and of degree 3 over $\mathbb{R}$. [Hint. Use Exercise 5.]
   (b) Now note that every cubic polynomial in $\mathbb{R}[X]$ has a zero in $\mathbb{R}$. (Proof via the Intermediate Value Theorem, as in [MS], page 103.)
   (c) Explain why (a) and (b) contradict the original supposition.

[Remarks: *elementary vector algebra.* The set of all vectors in the plane can be regarded as the set $\mathbb{R}^2$ of all ordered pairs of real numbers (a, b). The vector space $\mathbb{R}^2$ can be made into a field by adopting the definition of multiplication from $\mathbb{C}$; thus

$$(a, b)(c, d) = (ac - bd, ad + bc).$$

Elements of the form $(a, 0)$ then form a "copy" of $\mathbb{R}$ in $\mathbb{R}^2$.

The set of vectors in three-dimensional space, however, behaves quite differently. This set can be regarded as the set $\mathbb{R}^3$ of all ordered triples of real numbers (a,b,c). Exercise 6 shows that it is `impossible` to make the vector space $\mathbb{R}^3$ into an extension field of $\mathbb{R}$, regardless of how we choose the vector multiplication in $\mathbb{R}^3$. Barehanded proofs of this result are included in [HH] and [RY], while Chapter 7 of [IH] contains a discussion of a more general result known as the Theorem of Frobenius.]

Additional Reading for Chapter 4

In Chapter 4 we developed techniques for proving that polynomials of degree 3 or less are irreducible over $\mathbb{Q}$. Example 31.6 in [JF] will show you the sorts of computations that polynomials of degree 4 and higher can involve. There are also somewhat general techniques which are not restricted to such small degree polynomials and which apply over fields other than $\mathbb{Q}$. See, for example, Section 31.2 of [JF], Sections 101–107 of [AC], Chapter 9 of [WG], Section 2.1 of [CH], and Section 2.6 of [LS].

The problem of deciding, in general, whether a polynomial is irreducible over a given field is a difficult one. New, rather deep techniques, amenable to computer use, have been developed recently, and much work is still being done on the problem. See, for example, the chapter "Factorization of polynomials" by E. Kaltofen in [CA].

You may now wish to refer to other treatments of irr$(\alpha,\mathbf{F})$ such as those given in Section 35 of [JF], Section 109 of [AC] and Section 2.2 of [CH]. Our Theorem 4.3.1 is the link between our treatment and that in these books.

When discussing irr$(\alpha,\mathbf{F})$, we have usually taken $\mathbf{F}$ to be a subfield of the complex number field $\mathbf{C}$ and α to be a complex number because this is all that our impossibility proofs in Chapter 5 require. You should note that nearly all of our results about irr$(\alpha,\mathbf{F})$ generalize easily to the case where $\mathbf{F} \subseteq \mathbf{E}$ is any tower of fields and $\alpha \in \mathbf{E}$. Results about irr$(\alpha,\mathbf{F})$ are stated in this more general form in [AC] and [JF], for example.

[CA] *Computer Algebra: Symbolic and Algebraic Computation*, ed. B. Buchberger, G.E. Collins, R. Loos and R. Albrecht, Springer, Vienna, 2nd edition, 1983.

[AC] A. Clark, *Elements of Abstract Algebra*, Wadsworth, Belmont, California, 1971.

[JF] J.B. Fraleigh, *A First Course in Abstract Algebra*, 3rd edition, Addison-Wesley, Reading, Massachusetts, 1982.

[WG] W.J. Gilbert, *Modern Algebra with Applications*, Wiley, New York, 1976.

[CH] C.R. Hadlock, *Field Theory and its Classical Problems*, Carus Mathematical Monographs, No. 19, Mathematical Association of America, 1978.

[HH] H. Holden, "Rings on $\mathbb{R}^2$", *Mathematics Magazine*, 62 (1989), 48–51.

[IH] I.N. Herstein, *Topics in Algebra*, Blaisdell, New York, 1964.

[LS] L.W. Shapiro, *Introduction to Abstract Algebra*, McGraw-Hill, New York, 1975.

[MS] M. Spivak, *Calculus*, W.A. Benjamin, New York, 1967.

[RY] R.M. Young, "When is $\mathbb{R}^n$ a field?", *The Mathematical Gazette*, 72 (1988), 128–129.

CHAPTER 5

Straightedge and Compass Constructions

In the previous chapters we developed the algebraic machinery for proving that the three famous geometric constructions are impossible. In this chapter we introduce some geometry and start to show the connection between algebra and the geometry of constructions.

We begin by showing you how to do some basic constructions with straightedge and compass, and then how to construct line segments whose lengths are products, quotients or square roots of ones already constructed.

Strict rules, passed down from the Greek geometers, must be followed in performing a construction. An important part of this chapter involves spelling out these rules very precisely. This leads to the notion of constructible number, on which the impossibility proofs in the next chapter are based.

We conclude the chapter by showing that there is a connection between geometric constructions and fields of the form $\mathbb{F}(\sqrt{\gamma})$.

5.1 Standard Straightedge and Compass Constructions

In this section we show you how to do some basic constructions with straightedge and compass. In each of these constructions, we are given some points and some lines passing through these points (and possibly some circles). We can construct new lines and circles using the straightedge and compass as described in 1 and 2 below.

1. The *straightedge* may be used to draw a new line, extended as far as we like, through any two points already in the figure.

2. The *compass* may be used to draw new circles in two ways.

    (a) Put the compass point on one point in the figure and the pencil on another such point and draw a circle or an arc of a circle.

    (b) Place the compass point and pencil as in (a) but then move the compass point to a third point in the figure before drawing the circle (or an arc) with this third point as centre.

    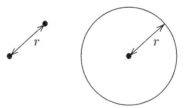

> We get new points where the new line or circle meets other lines or circles.

In the rest of this section we show how you can use a straightedge and compass to perform several constructions. These constructions will be used later as building blocks for other constructions.

Geometrical Constructions

5.1.1 Bisecting a line segment.

Purpose. *To construct the midpoint C of a given line segment AB.*

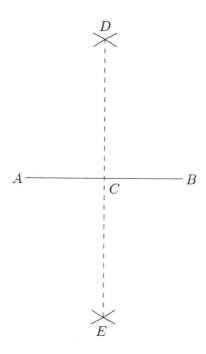

Method.

(i) Put the compass point on A and extend the compass until its pencil is exactly on B. Now draw an arc above AB and another arc below AB.

(ii) Put the compass point on B and extend the compass until its pencil is exactly on A. Now draw arcs to intersect the arcs in (i). Call the points of intersection D and E respectively.

(iii) Join D and E with the straightedge. Where DE meets AB is the required point C. ∎

[That C is the midpoint of AB seems obvious from the symmetry of the figure, and we give no formal proof of this (or of the validity of the constructions in the rest of this section). Readers familiar with Euclidean-geometry proofs may be interested in providing their own proofs, using facts about congruent triangles.]

5.1.2 Transferring a length (using a compass).

Purpose. Given a line segment AB and a (longer) line segment CD, to construct a point P on CD such that the line segments AB and CP have equal lengths.

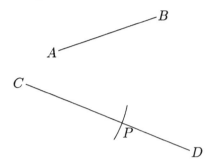

Method.

(i) Put the compass point on A and extend the compass until its pencil is exactly on B.

(ii) Put the compass point on C and with the same radius as in (i), draw an arc to cut CD at P. Then P is the required point. ∎

5.1.3 Bisecting an angle.

Purpose. To construct a line OC which bisects a given angle AOB.

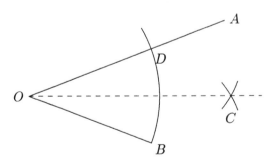

Method.

(i) With centre O and radius OB, draw an arc to meet OA (extended if necessary) at D.

(ii) Draw arcs with centres D and B, and radius BD, to meet at a point C. Then OC bisects angle AOB. ∎

5.1.4 Constructing an angle of 60°.

Purpose. Given a line segment OA, to construct a line OB so that angle $AOB = 60°$.

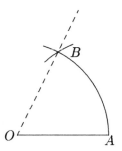

Method.

Draw arcs with centres O and A, and radius OA, which meet at a point B. Then angle $AOB = 60°$. ∎

5.1.5 Constructing an angle of 90°.

Purpose. Given a line segment OA, to construct a line OB so that angle $AOB = 90°$.

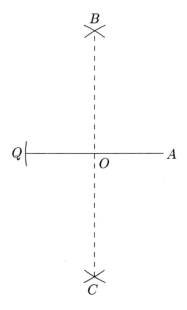

Method.

(i) With centre O draw an arc of radius OA which meets AO extended, at the point Q.

(ii) With centre A and radius AQ, draw arcs, one above the segment AQ and the other below AQ.

(iii) With centre Q and the same radius as in (ii), draw arcs to meet those in (ii) at points B and C respectively.

(iv) Join B and C, using the straightedge. This line passes through O and angle $AOB = 90°$. ∎

[Note that steps (ii)–(iv) are the same as those for bisecting the line segment AQ.]

5.1.6 Copying an angle.

Purpose. Given an angle ABC and a line segment DE, to construct a line EF so that the angles ABC and DEF are equal.

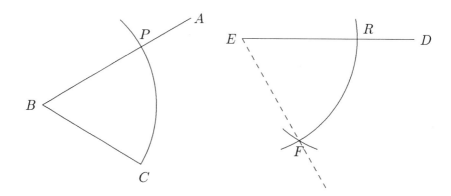

Method.

(i) With centre B and radius BC draw an arc to meet AB (extended if necessary) at P.

(ii) With centre E and the same radius as in (i), draw an arc to meet ED (extended if necessary) at R.

(iii) With centre R and the radius PC, draw an arc to meet the arc drawn in (ii) at F. Then angles ABC and DEF are equal. ∎

Geometrical Constructions

5.1.7 Constructing a line parallel to a given line.

Purpose. *Given points A and B and a point C not on the line through A and B, to construct a line segment CD which is parallel to AB.*

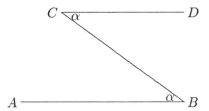

Method.

(i) Use the straightedge to draw the line joining C to B.

(ii) Use the angle copying construction given in 5.1.6 above to construct CD so that angles ABC and DCB are equal. Then CD is parallel to AB. ∎

5.1.8 Use of Compass and Straightedge.

In all of the above constructions, the straightedge has been used in strict conformity with the rules laid down by the ancient Greeks, who imagined the straightedge as being free of any markings. Thus, although it could be used to draw straight lines, it could not be used to measure distances between points or to transfer lengths.

Our use of the compass, on the other hand, calls for a more versatile compass than that allowed by the Greeks. The extra versatility lies in our assumption that the compass can be used to draw circles in <u>both</u> of the ways described in 2(a) and 2(b) at the start of Section 5.1, whereas the compass imagined by the Greeks could only be used in the <u>first</u> of these two ways. By allowing 2(b), we are assuming that our compass can be opened to a specified radius (as determined by the two points) and then picked up from the paper and used to draw a circle of this radius around a third point. We have used the compass this way in steps (i) and (ii) of 5.1.2 above, for example. The Greeks presumably thought of their compass as "collapsing" as soon as it was lifted from the paper and so it could not be used directly to transfer lengths.

It turns out, however, that it does not matter whether we use a *noncollapsing* compass (as in the constructions 5.1.2 and 5.1.6 above) or the apparently less powerful *collapsing* compass of the Greeks. It can be shown that any construction which can be performed with a noncollapsing compass can be performed also with a collapsing compass. Use of the collapsing compass may, of course, involve taking more steps in the construction. For the details, see Exercise 10 below.

──────────── **Exercises 5.1** ────────────

1. Given two points A and B and a point C not on the line through A and B, show how to construct (using only straightedge and compass) a line through C which is perpendicular to AB.

2. (a) On a sheet of paper construct a line segment $A'B'$ having exactly the same length as AB shown below:

   (b) Construct a line segment having one quarter of this length. For which positive integers n can you construct, in a similar way, a line segment whose length is $\frac{1}{n}$ times that of AB?

3. (a) On a sheet of paper construct an angle $A'O'B'$ which is equal to the angle AOB shown below:

   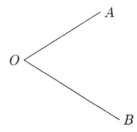

(b) Construct an angle which is one quarter of this angle.

(c) For which positive integers n can you construct, in a similar fashion, an angle which is $\frac{1}{n}$ times the angle AOB?

4. (a) Construct an equilateral triangle using straightedge and compass.

   (b) Which angle have you trisected in part (a)? How would you trisect an angle of 90°?

   (c) Can you trisect, with straightedge and compass, an angle of 60°? [Hint. Re-read Section 0.3.]

5. (a) Draw on a sheet of paper a line segment about 2 cm long and, taking this as your unit of length, construct line segments of lengths $\sqrt{3}$ and $1+\sqrt{3}$. [Hint. What is $\sin(60°)$?]

   (b) Appearances can be deceiving! Show that
   $$\sqrt{4+2\sqrt{3}} = 1+\sqrt{3},$$
   $$\sqrt{3+\sqrt{13+4\sqrt{3}}} = 1+\sqrt{3}.$$
   [Thus you have already constructed in part (a) line segments having these horrible looking lengths.]

6. How would you construct a line segment of length $\sqrt{2}$?

7. Find positive integers a, b, c, d (different from those in Exercise 5(b)) such that $\sqrt{a+b\sqrt{3}} = c+d\sqrt{3}$.

8. (a) Given an angle AOB, describe how you would construct (using only straightedge and compass) an angle which is approximately equal to one third of angle AOB, with an error which is less than one eighth of the original angle. [Hint. Find an integer a with the fraction $a/8$ sufficiently close to $1/3$.]

   (b) As for (a), but now suppose that the error is to be less than one sixteenth of the original angle.

   (c) Assume now that k is a positive integer. Describe a method of approximately trisecting an angle, using only straightedge and compass, which gives an error of at most 2^{-k} of the original angle.

9. Verify that the constructions 5.1.1, 5.1.3 and 5.1.4 can be performed exactly as described, using a collapsing compass.

10. First read through the construction set out below and then do (a) and (b) below.
    (a) Prove that the circle constructed at step (iii) has the required properties (thereby proving that the construction achieves its purpose).
    (b) Explain why the construction shows that a collapsing compass can be used to draw any circle that a noncollapsing compass can draw.

Purpose. Given points O, A and B, to construct a circle with centre O and radius AB with a `collapsing` compass.

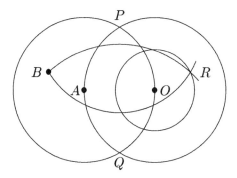

Method.

(i) Draw one circle with centre O and radius equal to the length of OA and another circle with centre A and radius OA. Let P and Q be the points where these two circles intersect.

(ii) Draw a circle with centre P and radius PB and another circle with centre Q and radius equal to QB. These circles intersect at B and also at another point which we call R. (In the diagram above, only parts of these circles are shown.)

(iii) Draw the circle with centre O and radius OR. This is the required circle.

11. The figure below shows how to construct a line segment CD which is parallel to a line AB by using a "rusty compass", that is, one which never alters its radius.

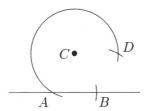

Why is CD parallel to AB?

(For more on this fascinating topic see Chapter 17 of [MG].)

5.2 Products, Quotients, Square Roots

5.2.1 Constructing a product.

Purpose. *Given line segments of lengths α and β, to construct a line segment of length $\alpha\beta$.*

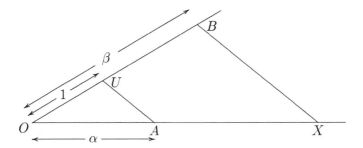

Method.

(i) Draw two lines which intersect in a single point O.

(ii) On one of these lines, construct a point A such that the length of OA is α.

(iii) On the other line construct

a point U such that the length of OU is 1, and

a point B such that the length of OB is β.

(iv) Construct (as in 5.1.7 above) a line through B parallel to UA and meeting the line OA, extended if necessary, at X.

Result. *The line segment OX has length $\alpha\beta$.*

Proof. Let x be the length of OX. Because the triangles OAU and OXB are similar,
$$\frac{x}{\alpha} = \frac{\beta}{1}$$
and so
$$x = \alpha\beta. \qquad \blacksquare$$

5.2.2 Constructing a quotient.

Purpose. *Given line segments of lengths α and $\beta \neq 0$, to construct a line segment of length α/β.*

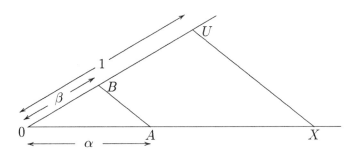

Method.

(i), (ii) and (iii) are as in 5.2.1.

(iv) Construct (as in 5.1.7) a line through U parallel to BA and meeting OA (extended, if necessary) at X.

Result. *The line segment OX has length α/β.*

Proof. Let x be the length of OX. By similar triangles,
$$\frac{x}{\alpha} = \frac{1}{\beta}$$
and so
$$x = \frac{\alpha}{\beta}. \qquad \blacksquare$$

5.2.3 Constructing square roots.

Purpose. *Given a line segment of length $\alpha > 0$, to construct a segment of length $\sqrt{\alpha}$.*

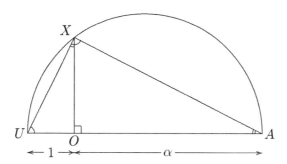

Method.

(i) Draw a line segment OA of length α and extend it backwards to U so that OU has length 1.

(ii) Bisect the line segment UA (as in 5.1.1) and construct a semicircle with the midpoint as centre and radius half the length of UA.

(iii) Erect a perpendicular to UA at O (as in 5.1.5) which intersects the semicircle at a point X, say.

Result. *The line segment OX has length $\sqrt{\alpha}$.*

Proof. Let x be the length of OX. The angle UXA is a right angle, being the angle subtended by a diameter of the circle at a point on the circumference. Hence the triangles UOX and XOA are similar. It follows that
$$\frac{x}{1} = \frac{\alpha}{x}$$
which gives
$$x = \sqrt{\alpha}. \qquad \blacksquare$$

──────────── Exercises 5.2 ────────────

1. (a) Start with a line segment which you define to have length 1 (a segment about 2 cm long should be ideal) and construct, with straightedge and compass, line segments of length

   (i) 4,   (ii) 1/3,   (iii) 7/5.

(b) Would you now revise your answer to Exercises 5.1 #2(b)?

(c) For which positive rational numbers r can you construct (with straightedge and compass) a line segment of length r?

2. Start with a line segment you define to have length 1 and then construct, with straightedge and compass, segments of length

   (i) $\sqrt{2}$,   (ii) $\sqrt{1+\sqrt{2}}$,   (iii) $\sqrt{2-\sqrt{2}}$,   (iv) $\sqrt[4]{5}$.

3. Suppose you are given a line segment of unit length. Convince yourself that you could construct (but do not bother to carry out the actual constructions accurately) line segments of lengths

   (i) $\sqrt{3/7 + 6\sqrt{5}}$,   (ii) $3\sqrt[4]{7/3} + 15/7$.

4. In the figure $|OA| = |OB| = 1$, angle $AOB = 36°$, r is the length of AB and the angles PAO and PAB are equal.

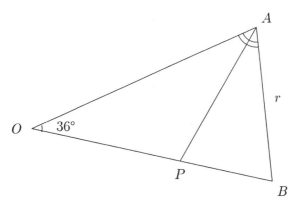

(a) Assume that you are given line segments of lengths 1 and r. Show how you would use straightedge and compass to construct an angle of 36°.

(b) By calculating the sizes of several angles in the figure, show that triangles OAB and APB are similar (isosceles) triangles and hence calculate r.

(c) Starting from a suitable line segment of unit length, construct (with straightedge and compass) a segment of length r.

5. Use Exercise 4 to construct (with straightedge and compass only) a regular pentagon (figure with 5 equal sides and angles) inscribed inside a circle. (Make the radius of the circle about 4 cm.)

Geometrical Constructions

5.3 Rules for Straightedge and Compass Constructions

A fairly precise description of the way the straightedge and compass can (and cannot) be used in performing constructions has already been given in the Introduction and, in more detail, at the start of Section 5.1. The purpose of this section is to spell out in complete detail the very precise rules (as passed down from the Greek geometers) which must be followed in these constructions.

In each of these construction problems, a set of points is given at the start, say
$$\{P_0, P_1, \ldots, P_m\},$$
which we shall call the *initial set* of points for the construction. Some lines and circles may also be given. Finitely many new points
$$P_{m+1}, P_{m+2}, \ldots, P_{m+n}$$
(and extra lines and circles) may be added, using straightedge and compass, to achieve the desired result.

We are not permitted to put points, lines and circles randomly on the page – rather we must base new lines and circles on points given at the start or already constructed. New points are where these lines and circles meet other lines and circles.

Thus the precise rules are that, at any stage of the construction process, we may add new points (and lines and circles) to the old ones already in existence by performing an operation of one of the following two types:

5.3.1 Construction Rules.

(i) *Draw a line through two old points P_i and P_j to get new points where this line, extended if necessary, intersects other lines and circles.*

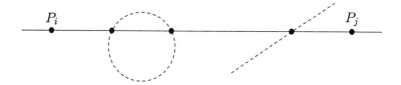

(ii) *Draw a circle with centre at an old point P_i and radius equal to the distance between two old points P_j and P_k to get new points where this circle intersects other lines and circles.*

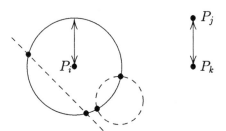

∎

Rule (i) tells us very precisely how the straightedge is to be used, while rule (ii) does the same for the compass. These rules tell us the only ways we are allowed to enlarge the set of points, lines and circles.

Note that as a special case, P_j may be the same as P_i in rule (ii). This permits us to draw a circle with centre at an old point P_i and radius equal to the distance between P_i and another old point P_k.

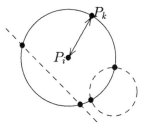

Thus rule (ii) does allow us to use the compass in the two ways 2(a) and 2(b) described at the start of Section 5.1. Case 2(a) is when $P_i = P_j$ and 2(b) is when P_i and P_j are different.

In summary then, in any construction problem,

> begin with an initial set of points (and lines and circles), and enlarge the set of points using finitely many operations of types (i) and (ii)

to achieve the desired result.

As you would expect, all of the constructions in Sections 5.1 and 5.2 have been carried out according to these rules. In Example 5.3.2

Geometrical Constructions

below, we spell this out for the construction in 5.1.1 and leave you (see Exercises 5.3 #1–3) to check this for the rest of the constructions.

5.3.2 Example. The construction "bisect a line segment" given in 5.1.1 follows the rules just given.

Proof. The initial set of points for the construction is $\{A, B\}$. Hence, in the notation above we may put $P_0 = A$ and $P_1 = B$.

The construction proceeds by the drawing of two circles with centres A and B respectively, and radius equal to the distance between the points A and B. Thus these circles are of the type permitted by rule (ii), and so the new points D and E, being points where the circles intersect, can be obtained by two operations of type (ii). Hence we may put $P_2 = D$ and $P_3 = E$.

The final stage in the construction involves drawing the line DE joining these two points to get the point C where this line intersects the line AB. Hence C can be obtained by an operation of type (i), and so we may choose $P_4 = C$. The desired midpoint C has thus been obtained from the set of initial points by successive operations of the types (i) and (ii). ∎

We conclude this section by re-examining the famous construction problems in the light of the construction rules given above.

5.3.3 Doubling the Cube. We are given a cube and asked to construct a cube with double the volume. We first translate this into the following equivalent problem in two dimensions. Given two points P_0 and P_1 whose distance apart is equal to one side of the original cube, we are asked to construct two points P_i and P_j whose distance apart is exactly $\sqrt[3]{2}$ times the distance between P_0 and P_1.

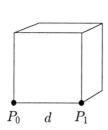
$P_0 \quad d \quad P_1$

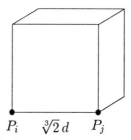
$P_i \quad \sqrt[3]{2}\,d \quad P_j$

The question is whether or not this is possible using operations of types (i) and (ii) only. ∎

5.3.4 Squaring the Circle. As in 5.3.3, this is equivalent to being given just two points P_0 and P_1 (distance apart equal to the radius of the circle) and being asked to construct two points P_i and P_j whose distance apart is exactly $\sqrt{\pi}$ times the distance between P_0 and P_1.

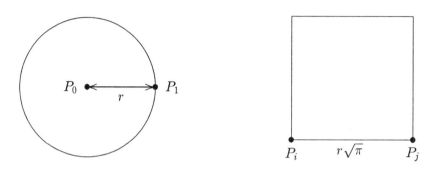

As always, the question is whether or not this is possible using only finitely many operations of types (i) and (ii). ∎

5.3.5 Trisecting an Angle. We are given the points P_0, P_1 and P_2, which determine the angle. We are then asked to construct (using only operations of types (i) and (ii)) a point P_i such that angle $P_i P_0 P_1$ is exactly one third of angle $P_2 P_0 P_1$.

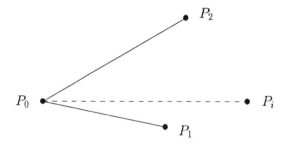

∎

——————————— **Exercises 5.3** ———————————

1. Verify that the construction 5.1.2 (transferring a length) can be performed by successive operations of types (i) and (ii) of 5.3.1.

Geometrical Constructions

2. Verify that the construction 5.1.3 (bisection of an angle) can be performed by successive operations of the types (i) and (ii).

3. Satisfy yourself that all the constructions in Sections 5.1 and 5.2 can be performed by successive operations of types (i) and (ii).

5.4 Constructible Numbers and Fields

In analysing construction problems, it often helps to introduce a convenient unit of length. For example, in analysing the problem of doubling the cube, a convenient unit of length is the length of one side of the given cube. Doubling the cube (see 5.3.3 above) is then possible if, given points P_0 and P_1 one unit apart, we can construct points P_i and P_j whose distance apart is $\sqrt[3]{2}$, using only operations of types (i) and (ii). This leads to the following definition.

5.4.1 Definition. Let γ be a real number with absolute value $|\gamma|$. Then γ is said to be *constructible* if we can construct points P_i and P_j whose distance apart is $|\gamma|$ units by: starting from an initial set of points $\{P_0, P_1\}$ whose distance apart is 1 unit and then performing a finite number of operations of types (i) and (ii) of 5.3.1. ∎

Clearly, doubling the cube is possible if and only if the number $\sqrt[3]{2}$ is constructible, while squaring the circle is possible if and only if $\sqrt{\pi}$ is constructible.

The following two theorems and the related discussion make it clear that the set of constructible numbers is quite large. They also mark the place where geometry and algebra begin to overlap.

5.4.2 Theorem. [Field with Square Roots Theorem] *The set CON of all constructible numbers is a subfield of $\mathbb{R}$. Furthermore*

(a) all rational numbers are in CON, and

(b) if $\alpha \in CON$ and $\alpha > 0$ then $\sqrt{\alpha} \in CON$.

Proof.

> We aim to show that CON is a subfield of $\mathbb{R}$; that is, the operations of addition, subtraction, multiplication and division (except by 0) can be performed without restriction in CON.

Let α and β be in CON. Hence line segments of lengths $|\alpha|$ and $|\beta|$ can be constructed by successive operations of types (i) and (ii) of 5.3.1, starting from a segment of length 1 unit.

Because we can transfer lengths (by 5.1.2), it is easy to see that segments of lengths $|\alpha + \beta|$ and $|\alpha - \beta|$ can be constructed from the above two segments. (If, for example, the segment AB has length α and segment CD has length β, we can extend AB and use 5.1.2 to construct a point E so that BE has length β and AE has the desired length $\alpha + \beta$.) We can also construct, using operations of types (i) and (ii), segments of lengths $|\alpha\beta|$ and $|\alpha/\beta|$ (if $\beta \neq 0$), by 5.2.1 and 5.2.2.

Thus the numbers $\alpha + \beta$, $\alpha - \beta$, $\alpha\beta$ and α/β (if $\beta \neq 0$) are all constructible and so are in CON. Hence CON is a field.

We now prove that all rational numbers are in CON.

Since we are given a segment of length 1, we can construct segments of lengths $1 + 1 = 2$, $2 + 1 = 3$ and so on, from which we see that we can construct segments of length equal to any positive integer. Hence, by 5.2.2, we can construct a segment of length equal to any desired positive rational number m/n (with $m, n \in \mathbb{N}$). Because of the use of $|\gamma|$ in Definition 5.4.1, it follows that all rational numbers (negative as well as positive) are in CON. (Alternatively, the result that $\mathbb{Q} \subseteq CON$ follows from Proposition 1.1.1 and the fact that CON is a field.)

Finally, by 5.2.3, if $\alpha \in CON$ and $\alpha > 0$, then $\sqrt{\alpha}$ is constructible. ∎

By applying this theorem repeatedly we see that each of the numbers

$$1 + 1 = 2, \quad 5, \quad \frac{2}{5}, \quad \frac{2}{5} + \sqrt{2}, \quad 3 + \sqrt{\frac{2}{5} + \sqrt{2}}, \quad 2 + \sqrt{3 + \sqrt{\frac{2}{5} + \sqrt{2}}}$$

is constructible.

Thus we see that every real number which is obtained from $\mathbb{Q}$ by performing successive field operations and taking square roots is constructible. The following theorem expresses this precisely, in the notation of field extensions.

Geometrical Constructions

5.4.3 Theorem. [Successive Square Roots Give Constructibles] *A real number γ is constructible if there exist positive real numbers $\gamma_1, \gamma_2, \ldots, \gamma_n$ such that*

$\gamma_1 \in \mathbb{F}_1$ *where* $\mathbb{F}_1 = \mathbb{Q}$,

$\gamma_2 \in \mathbb{F}_2$ *where* $\mathbb{F}_2 = \mathbb{F}_1(\sqrt{\gamma_1})$,

$\vdots \qquad\qquad \vdots$

$\gamma_n \in \mathbb{F}_n$ *where* $\mathbb{F}_n = \mathbb{F}_{n-1}(\sqrt{\gamma_{n-1}})$,

and, finally,

$\gamma \in \mathbb{F}_{n+1}$ *where* $\mathbb{F}_{n+1} = \mathbb{F}_n(\sqrt{\gamma_n})$.

Proof. This follows immediately from the Field with Square Roots Theorem. ∎

Notice that there is a tower of fields

$$\mathbb{Q} \subseteq \mathbb{F}_1 \subseteq \mathbb{F}_2 \subseteq \ldots \subseteq \mathbb{F}_n \subseteq \mathbb{F}_{n+1}$$

implicit in this theorem. This tower will play an important rôle in the impossibility proofs in the next chapter.

In the following example, the fields $\mathbb{F}_1, \mathbb{F}_2, \ldots$ and the numbers $\gamma_1, \gamma_2, \ldots$ relevant to a particular γ are indicated.

5.4.4 Example. Let

$$\gamma = 5\sqrt{2} + \frac{\sqrt{8 - 3\sqrt{2}}}{(1 - \sqrt{2})}.$$

Prove that γ is constructible by using the "Successive Square Roots Give Constructibles" Theorem.

Proof.

> We must produce a positive integer n and positive real numbers $\gamma_1, \gamma_2, \ldots, \gamma_n$ as in the statement of the "Successive Square Roots Give Constructibles" Theorem such that $\gamma \in \mathbb{F}_{n+1} = \mathbb{F}_n(\sqrt{\gamma_n})$.

Put $\gamma_1 = 2 \in \mathbb{F}_1$ where $\mathbb{F}_1 = \mathbb{Q}$,

$\gamma_2 = 8 - 3\sqrt{2}$

$ = 8 - 3\sqrt{\gamma_1} \in \mathbb{F}_2$ where $\mathbb{F}_2 = \mathbb{F}_1(\sqrt{\gamma_1})$.

Hence $\gamma = 5\sqrt{\gamma_1} + \dfrac{\sqrt{\gamma_2}}{1 - \sqrt{\gamma_1}} \in \mathbb{F}_3$ where $\mathbb{F}_3 = \mathbb{F}_2(\sqrt{\gamma_2})$.

Thus we have produced positive real numbers γ_1, γ_2 which satisfy the hypothesis of the "Successive Square Roots Give Constructibles" Theorem. Hence γ is constructible. ∎

In this chapter we have concentrated on showing that there are many different constructible numbers. In the next chapter (where we prove the impossibility of the three famous constructions) we take the opposite point of view and concentrate on showing that there are many real numbers which are not constructible. Indeed, we will prove there that the only constructible numbers are those given by the

"Successive Square Roots Give Constructibles" Theorem.

That is, we prove that any number which cannot be obtained from $\mathbb{Q}$ by performing successive field operations and taking square roots is not constructible. This result, which we call the

"All Constructibles Come From Square Roots" Theorem,

is the converse of Theorem 5.4.3.

────────────── Exercises 5.4 ──────────────

1. Let $\gamma = 3\sqrt{2} + \sqrt{5 - 3\sqrt{2}}$.

    Prove in the following two different ways that γ is constructible:

    (a) firstly, by applying the Field with Square Roots Theorem,

    (b) secondly, by applying the "Successive Square Roots Give Constructibles" Theorem.

2. Show that the zeros in $\mathbb{R}$ of the polynomial $2X^4 - 6X^2 - 3$ are all constructible real numbers. [Hint. This is a quadratic in X^2.]

3. Given constructible numbers α and β, describe how you would use 5.1.2 to construct a segment of length $|\alpha + \beta|$ in each of the following cases:

    (i) α and β have the same sign,

    (ii) α and β have opposite sign.

4. [This exercise shows that, in some sense, if we could use infinitely many operations of types (i) and (ii), every real number would be constructible.]

(a) Show that every real number can be approximated arbitrarily closely by a rational number.

(b) Show that every real number can be approximated arbitrarily closely by a constructible number.

Additional Reading for Chapter 5

Basic straightedge and compass constructions, including bisection of line segments and angles, are given in Chapter 4, Lesson 8 of [HJ]; proofs that the constructions achieve their intended purpose are included. A discussion of collapsing and modern (or noncollapsing) compasses is given in Section 4.1 of [HE]. The constructions for sums, differences, products, quotients and square roots of constructible numbers are given in Section 39.1 of [JF]. The rules governing the use of straightedge and compass are explained on pages 7 and 8 of [EH].

Those who have previously met the constructions of bisecting a line segment, bisecting an angle, etc, will notice that our constructions do not use arbitrary choices. Phrases which are common in elementary geometry textbooks such as "draw a *suitable* arc" or "choose a *suitable* radius" are absent from our presentation. For a discussion of such indeterminate constructions see page 13 of [IHH].

It is interesting to note that all constructions that can be performed using straightedge and compass can also be performed using a compass alone. It is also true that all such constructions can be performed using a straightedge and a single circle, or even an arc of a circle, together with its centre. For a discussion of this work of Jean Victor Poncelet, Jacob Steiner, Lorenzo Mascheroni and Georg Mohr see Chapter 17 of [MG], and also [AK] and [JS].

A very different problem from that considered in this book is that of squaring the circle using scissors. More precisely, the problem is to cut a circle (or rather, a disc) into finitely many pieces using scissors (cutting along Jordan arcs) and then to reassemble the pieces into a square of the same area. It is proved in [LD], however, that this construction is impossible. Whether the construction is possible when more weird "pieces" are allowed than those obtained by cutting with scissors is described in [SW] as an unsolved problem. As mentioned in [RR] this problem has been solved recently with a proof that the construction is possible. The analagous problem in three (or higher) dimensions is solved by the Banach-Tarski (Paradox) Theorem. An exposition of this astonishing theorem appears in [RF].

[LD] L. Dubins, M.W. Hirsch and J. Karush, "Scissor Congruence", *Israel Journal of Mathematics*, 1 (1963), 239–247.

[HE] H. Eves, *A Survey of Geometry*, Allyn and Bacon, Boston, Massachusetts, 1972.

[JF] J.B. Fraleigh, *A First Course in Abstract Algebra*, 3rd edition, Addison-Wesley, Reading, Massachusetts, 1982.

[RF] R.M. French, "The Banach-Tarski Theorem", *The Mathematical Intelligencer*, 10 (1988), 21–28.

[MG] M. Gardner, *Mathematical Circus*, Penguin Books, Middlesex, England, 1981.

[EH] E.W. Hobson, *Squaring the Circle*, Cambridge University Press, 1913; reprinted in *Squaring the Circle, and Other Monographs*, Chelsea, 1953.

[HH] H.P. Hudson, *Ruler and Compass*, Longmans Green, 1916; reprinted in *Squaring the Circle, and Other Monographs*, Chelsea, 1953.

[HJ] H.R. Jacobs, *Geometry*, Freeman, San Francisco, California, 1974.

[AK] A.N. Kostovskii, *Geometrical Constructions Using Compasses Only*, Blaisdell, New York, 1961.

[RR] R. Ruthen, "Squaring the Circle", *Scientific American*, 261 (1989), 11–12.

[JS] J. Steiner, "Geometrical Constructions with a Ruler Given a Fixed Circle with its Center", *Scripta Mathematica*, 14 (1948), 187–264.

[SW] S. Wagon, "Circle-Squaring in the Twentieth Century", *The Mathematical Intelligencer*, 3 (1981), 176–181.

CHAPTER 6

Proofs of the Impossibilities

Within this chapter it finally becomes clear why the three famous geometrical constructions are impossible. You are now, at long last, in a position to see the solutions of problems which defied the world's best mathematicians for over two thousand years. The key to the solutions lies in combining the geometrical ideas from Chapter 5 with the algebraic ideas from earlier chapters.

Our first task is to show that there are no constructible numbers other than those already found in Chapter 5. This enables us to show that every constructible number is algebraic over $\mathbb{Q}$ and has a degree which is a power of 2.

We are then able to show that, if doubling the cube or trisecting an arbitrary angle were possible, there would have to exist a constructible number whose degree is not a power of 2. This contradiction means that we can be certain that these constructions (Problems I and II) are impossible.

We complete the proof of the impossibility of squaring the circle (Problem III) by showing in Chapter 7 that π is not algebraic over $\mathbb{Q}$.

6.1 Non-Constructible Numbers

In Chapter 5 we found that real numbers like

$$2 + \sqrt{3 + \sqrt{\frac{2}{5} + \sqrt{2}}},$$

which are obtained from elements of $\mathbb{Q}$ by successively applying field operations and taking square roots, are all constructible. For the purposes of this chapter, however, it is of the utmost importance to know whether we can go in the reverse direction: *can every constructible number be expressed in terms of repeated square roots and field operations, starting from numbers in $\mathbb{Q}$?*

The following theorem shows that the answer is "yes".

6.1.1 Theorem. [All Constructibles Come From Square Roots] *If the real number γ is constructible, then there exist positive real numbers $\gamma_1, \gamma_2, \ldots, \gamma_n$ such that*

$\gamma_1 \in \mathbb{F}_1$ *where* $\mathbb{F}_1 = \mathbb{Q}$,

$\gamma_2 \in \mathbb{F}_2$ *where* $\mathbb{F}_2 = \mathbb{F}_1(\sqrt{\gamma_1})$,

$\vdots \qquad\qquad \vdots$

$\gamma_n \in \mathbb{F}_n$ *where* $\mathbb{F}_n = \mathbb{F}_{n-1}(\sqrt{\gamma_{n-1}})$,

and, finally,

$\gamma \in \mathbb{F}_{n+1}$ *where* $\mathbb{F}_{n+1} = \mathbb{F}_n(\sqrt{\gamma_n})$.

Proof. We postpone this rather long proof until Section 6.3. The idea of the proof is that when you intersect lines and circles, the worst that can happen is that you need extra square roots to describe the coordinates of the points of intersection. The example below shows this happening in a particular case. ∎

6.1.2 Example. Start from two points $P_0 = (0,0)$ and $P_1 = (1,0)$, at a distance 1 apart, as in the definition of a constructible number (Definition 5.4.1). Draw two circles with centres at P_0 and P_1 respectively, each with radius P_0P_1. Show that the coordinates of each of the points of intersection, P_2 and P_3, of the two circles involve a single square root.

Proofs of the Impossibilities

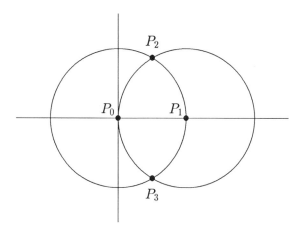

Proof. The circles have equations

$$x^2 + y^2 = 1, \tag{1}$$
$$(x-1)^2 + y^2 = 1. \tag{2}$$

The points of intersection satisfy both equations. Subtracting (1) from (2) gives $x = 1/2$ and substituting in (1) gives $y = \pm\sqrt{3}/2$. Thus the two points of intersection are $P_2 = \left(\frac{1}{2}, \frac{\sqrt{3}}{2}\right)$ and $P_3 = \left(\frac{1}{2}, -\frac{\sqrt{3}}{2}\right)$, which involve nothing worse than a square root, $\sqrt{3}$. ∎

We now apply the "All Constructibles Come From Square Roots" Theorem to show that every constructible number must be algebraic over $\mathbb{Q}$ and must have degree over $\mathbb{Q}$ which is a power of 2. This result, which is the key to the impossibility proofs, enables us to be certain that many numbers are **not** constructible.

6.1.3 Theorem. [Degree of a Constructible Number Theorem]
If a real number γ is constructible, then γ is algebraic over $\mathbb{Q}$ and $\deg(\gamma, \mathbb{Q}) = 2^s$ for some integer $s \geq 0$.

Proof. Let γ be constructible and let $\gamma_1, \ldots, \gamma_n$ be as in the "All Constructibles Come From Square Roots" Theorem. The number $\sqrt{\gamma_i}$ is a zero of the polynomial $X^2 - \gamma_i$, which is in $\mathbb{F}_i[X]$ since $\gamma_i \in \mathbb{F}_i$. Hence, by Definition 2.3.2,

$$\deg(\sqrt{\gamma_i}, \mathbb{F}_i) = 1 \text{ or } 2$$

and since $\mathbb{F}_{i+1} = \mathbb{F}_i(\sqrt{\gamma_i})$ it follows from Theorem 3.2.4 that

$$[\mathbb{F}_{i+1} : \mathbb{F}_i] = 1 \text{ or } 2 \qquad (1 \leq i \leq n).$$

Repeated application of the Dimension for a Tower Theorem (Theorem 3.4.3) to the tower

$$\mathbb{Q} = \mathbb{F}_1 \subseteq \mathbb{F}_2 \subseteq \mathbb{F}_3 \subseteq \ldots \subseteq \mathbb{F}_{n+1}$$

shows that

$$[\mathbb{F}_{n+1} : \mathbb{Q}] = [\mathbb{F}_{n+1} : \mathbb{F}_n][\mathbb{F}_n : \mathbb{F}_{n-1}] \ldots [\mathbb{F}_2 : \mathbb{F}_1]$$
$$= 2^u, \quad \text{for some integer } u \geq 0.$$

It follows from Theorem 4.4.1 that γ is algebraic over $\mathbb{Q}$. Also, by considering the tower

$$\mathbb{Q} \subseteq \mathbb{Q}(\gamma) \subseteq \mathbb{F}_{n+1}$$

we see that $\deg(\gamma, \mathbb{Q})$ is a factor of $[\mathbb{F}_{n+1} : \mathbb{Q}]$. Hence

$$\deg(\gamma, \mathbb{Q}) = 2^s$$

for some integer $s \geq 0$. ∎

It should be noted, perhaps with a little surprise, that the *converse* of the above result is false. A counterexample to the converse is given as Example 13-18 in [WG]; this example describes a specific number γ which is not constructible but which is algebraic over $\mathbb{Q}$ with $\deg(\gamma, \mathbb{Q}) = 2^2$.

─────────────── **Exercises 6.1** ───────────────

1. Let $\gamma = 2 + \sqrt{3 + \sqrt{5}}$.

   (a) Prove that γ is constructible by using the Field with Square Roots Theorem (Theorem 5.4.2).

   (b) Prove that γ is constructible by using the "Successive Square Roots Give Constructibles" Theorem (Theorem 5.4.3).

   [Hint. Show how to choose γ_1, γ_2, etc.]

   (c) What does the Degree of a Constructible Number Theorem now tell you about the degree of γ?

2. Combine the "Successive Square Roots Give Constructibles" and the "All Constructibles Come From Square Roots" theorems into one theorem.

3. Use the Degree of a Constructible Number Theorem and the fact that
$$\mathrm{irr}(\sqrt[3]{5}, \mathbb{Q}) = X^3 - 5$$
to prove that $\sqrt[3]{5}$ is not constructible.

4. (a) Use the Degree of a Constructible Number Theorem to prove that the number $1 + \sqrt[3]{5}$ is not constructible. Include a full proof that $\deg(1 + \sqrt[3]{5}, \mathbb{Q})$ is what you claim it to be.

   (b) Use the result of Exercise 3 and the fact that the constructible numbers form a field to give an alternative proof that the number $1 + \sqrt[3]{5}$ is not constructible.

5. In Example 6.1.2 in the text, let P_4 be the point of intersection of the line segments P_0P_1 and P_2P_3.

   (a) Write down the equation of the circle with centre P_2 and radius P_2P_4.

   (b) Show that this new circle intersects the two circles already constructed in four points all of whose coordinates lie in the field $\mathbb{Q}(\sqrt{3})(\sqrt{39})$.

6.** Show that the the set of all constructible numbers is countably infinite.

   [Hint. Recall that a set S is said to be *countably infinite* if there exists a function $f : \mathbb{N} \to S$ which is one-to-one and onto.]

6.2 The Three Constructions are Impossible

At long last, we are in a position to see why the constructions in the Introduction are impossible. In each case we shall show that if the construction were possible, then there would exist a constructible number which is not algebraic or whose degree over $\mathbb{Q}$ is not a power of 2. Because the existence of such a constructible number contradict the Degree of a Constructible Number Theorem ("
6.1.3), we can be certain that each construction is impossib'

6.2.1 Problem I - Doubling the cube.

A cube whose volume is 2 must have sides whose lengths are $\sqrt[3]{2}$. Hence doubling the cube amounts to constructing a line segment whose length is $\sqrt[3]{2}$ (starting from a line segment of length 1 and using only straightedge and compass). Thus if the cube could be doubled, then $\sqrt[3]{2}$ would be a constructible number.

By Example 4.3.3, however,

$$\mathrm{irr}(\sqrt[3]{2}, \mathbb{Q}) = X^3 - 2$$

and so

$$\deg(\sqrt[3]{2}, \mathbb{Q}) = 3,$$

which is not a power of 2. This shows that $\sqrt[3]{2}$ is not a constructible number, and so the cube cannot be doubled.

6.2.2 Problem II - Trisecting an arbitrary angle.

If we could trisect every angle then, in particular, we could trisect an angle of 60°. But since 60° is an angle which can be constructed (see Section 5.1), it follows that we could then construct an angle of 20° also. Hence, using a right-angled triangle (see Exercises 6.2 #2 below), we could also construct a line segment of length cos(20°).

It is easy to show, however, that the irreducible polynomial of cos(20°) over $\mathbb{Q}$ is a cubic (see Exercises 6.2 #3 below for the details). Hence

$$\deg(\cos(20°), \mathbb{Q}) = 3$$

and so cos(20°) is not constructible, as 3 is not a power of 2. Thus it is not possible to construct an angle of 20°.

You should note that the argument just given shows that there is one angle (namely 60°) which cannot be trisected. But it is easy to deduce from this result (as, for example, in Exercises 6.2 #4 below) that there are infinitely many angles which cannot be trisected.

Note also that the argument given above that the angle 60° cannot be trisected, relies crucially on the fact that we can construct an angle of 60°. Indeed, the argument given cannot be used to decide whether it is possible to trisect other angles (such as 10°) which cannot themselves be constructed (starting just from a pair of points P_0 and P_1).

6.2.3 Problem III - Squaring the circle.

A circle of radius 1 has area equal to π square units, and so a square with the same area would have sides of length $\sqrt{\pi}$. Thus if squaring the circle could be done with straightedge and compass, it would follow that $\sqrt{\pi}$ is constructible. This would mean (by Theorem 6.1.3) that $\sqrt{\pi}$ is algebraic over $\mathbb{Q}$, and hence that $\pi = \sqrt{\pi}.\sqrt{\pi}$ is algebraic over $\mathbb{Q}$ by Theorem 4.4.4. We complete the proof of the impossibility of squaring the circle in Chapter 7 in which we show that π is transcendental (that is, not algebraic over $\mathbb{Q}$).

6.2.4 Other constructions which are impossible.

In [CH] it is shown that not every angle can be divided into *five* equal parts, with straightedge and compass. As a generalization, we indicate a proof below (in Exercises 6.2 #7 and 8) that, for any given positive integer n which is not a power of 2, it is impossible to divide an arbitrary angle into n equal parts.

Although there are proofs for the impossibility of trisecting angles which do not involve the full algebraic machinery developed in this book, it is remarked in [CH] that such proofs are not so easy to adapt to proving further impossibilities such as that of quintisecting an arbitrary angle.

─────────────── Exercises 6.2 ───────────────

1. Decide in each case whether the number is constructible and give reasons. You may assume $\sqrt[3]{2}$ is not constructible.
   (i) $\sqrt{2+\sqrt{2}}$,   (ii) $\sqrt[4]{2}$,   (iii) $\sqrt[6]{2}$,
   (iv) $\sqrt{2}\sqrt[3]{2}$,   (v) $\sqrt{2}+\sqrt[3]{2}$.

2. (a) Assume that an angle θ has been constructed. Show how you would construct lengths $\sin\theta$ and $\cos\theta$ from it, using straightedge and compass.
   (b) Assume, conversely, that a length $\sin\theta$ has been constructed.
     (i) Show *algebraically* that $\cos\theta$ is then constructible.
        [Hint. Use the Field with Square Roots Theorem.]
     (ii) Show *geometrically* how to construct $\cos\theta$.

(iii) Show how you would use straightedge and compass to construct the angle θ $(0 < \theta < \frac{\pi}{2})$ from $\sin\theta$ and $\cos\theta$.

3. (a) Use the identity $\cos(3\theta) = -3\cos\theta + 4\cos^3\theta$ to show that $\cos(20°)$ is a zero of the polynomial $8X^3 - 6X - 1$.

   (b) Deduce that $\cos(20°)$ is not a constructible number. Which part of Exercise 2 now shows that an angle of 20° cannot be constructed with straightedge and compass?

   (c) Which angle have you proved cannot be trisected?

4. (a) Show that an angle of 30° cannot be trisected.
   [Hint. Note that an angle of 60° cannot be trisected.]

   (b) Repeat for an angle of 15°.

   (c) Show that there are infinitely many angles which cannot be trisected.

5. (a) Let $\alpha \in \mathbb{R}$ be a zero of the polynomial $p(X) = X^3 + 2X + 1$ in $\mathbb{Q}[X]$. Prove each of the following:

   (i) $p(X)$ is irreducible over $\mathbb{Q}$.

   (ii) $p(X) = \mathrm{irr}(\alpha, \mathbb{Q})$.

   (iii) α is not constructible.

   (b) Can we assert that $1 - \alpha + \alpha^2 \neq 0$? Why?

6. Let α and $\beta \in \mathbb{R}$. In each case decide whether the sentence is true or false. Give reasons.

   (a) If α and β are constructible then so is $\alpha + \beta$.

   (b) If $\alpha + \beta$ is constructible then so are α and β.

   (c) If $\alpha\beta$ is constructible then so are α and β.

   (d) If $\alpha + \beta$ and $\alpha\beta$ are both constructible then so are α and β.
   [Hint for (d). Can you express α and β in terms of $\alpha + \beta$ and $\alpha\beta$?]

7.* In this exercise you will prove the Cos($n\theta$)-formula: *For each prime number $n > 2$ there are integers $a_1, a_3, \ldots, a_{n-2}, a_n$ such that*

$$\cos(n\theta) = a_1\cos\theta + a_3\cos^3\theta + \ldots + a_{n-2}\cos^{n-2}\theta + a_n\cos^n\theta$$

for all $\theta \in \mathbb{R}$, where $a_n = 2^{n-1}$ and the remaining coefficients $a_1, a_3, \ldots, a_{n-2}$ all have n as a factor.

(a) Using the De Moivre Theorem (see [WL]) and the Binomial Theorem show that, if we let $c = \cos\theta$, then

$$\cos(n\theta) = c^n + \binom{n}{2}c^{n-2}(c^2-1) + \binom{n}{4}c^{n-4}(c^2-1)^2 + \ldots + \binom{n}{n-1}c(c^2-1)^{\frac{n-1}{2}}.$$

The following parts of this exercise assume that the right hand side of this formula is rearranged to give a polynomial in c.

(b) Show that the coefficient of c^n in this polynomial is

$$1 + \binom{n}{2} + \binom{n}{4} + \ldots + \binom{n}{n-1}$$

and then use the Binomial Theorem to show that this equals

$$\frac{1}{2}\left((1+1)^n + (1+(-1))^n\right) = 2^{n-1}.$$

(c) By evaluating both sides of the equation in (a) at suitable θ, show that the constant term of the polynomial is zero.

(d) State why the prime n is a factor of $\binom{n}{r}$ for $1 \leq r \leq n-1$. Deduce that the coefficients of the remaining powers of c in the polynomial in part (a) all have n as a factor.

8.* **Eisenstein's Irreducibility Criterion** states : *the polynomial*

$$a_0 + a_1 X + \ldots + a_{n-1}X^{n-1} + a_n X^n \in \mathbb{Z}[X]$$

is irreducible over $\mathbb{Q}$ if there exists a prime number p such that p is not a factor of a_n, but p is a factor of $a_0, \ldots, a_{n-1}$, and p^2 is not a factor of a_0. (For a proof see [JF], Theorem 31.4.)

(a) Let $n > 2$ be a prime number. Choose θ so that $\cos(n\theta) = \frac{n}{n+1}$ and then deduce from the $\cos(n\theta)$-formula and the Eisenstein Irreducibility Criterion that $\deg(\cos\theta, \mathbb{Q}) = n$.

(b) Deduce that, for each positive integer n which is not a power of 2, there exists an angle which cannot be divided into n equal parts via straightedge and compass constructions.

6.3 Proving the "All Constructibles Come From Square Roots" Theorem

In order to describe the points (and the associated lines and circles) which appear when the operations (i) and (ii) of 5.3.1 are applied, we work with the coordinates of these points. To do so, we introduce the following definitions.

6.3.1 Definitions. Let F be a subfield of R.

A point is an F-*point* if both of its coordinates are in F.

A line is an F-*line* if it passes through two F-points.

A circle is an F-*circle* if its centre is an F-point and its radius is the distance between two F-points. ∎

6.3.2 Example.

(a) $(2,1)$ is a Q-point, because both its coordinates belong to Q.

(b) $(1, \sqrt{2})$ is a $Q(\sqrt{2})$-point.

(c) $\{(x,y) : y = x\}$ is a Q-line, because it contains the two Q-points $(0,0)$ and $(1,1)$.

(d) $\{(x,y) : y = \sqrt{2}x\}$ is a $Q(\sqrt{2})$-line but not a Q-line.

(e) $\{(x,y) : x^2 + y^2 = 4\}$ is a Q-circle, because it has the Q-point $(0,0)$ as its centre while its radius is the distance between the Q-points $(0,0)$ and $(0,2)$.

(f) $\{(x,y) : x^2 + y^2 = 2\}$ is a $Q(\sqrt{2})$-circle. ∎

If we intersect a Q-line with a Q-circle, do we always get a Q-point? The following example shows that the answer is "no".

6.3.3 Example. The Q-line $\{(x,y) : y = x\}$ intersects the Q-circle $\{(x,y) : x^2 + y^2 = 4\}$ at the two points $(\sqrt{2}, \sqrt{2})$ and $(-\sqrt{2}, -\sqrt{2})$. These points are not Q-points. However, they are $Q(\sqrt{2})$-points. ∎

6.3.4 Remarks.
Assume that we are performing a construction as described in Section 5.3 and that we have already obtained points $P_0, P_1, \ldots, P_k$.

Let $\mathbb{F}$ be a subfield of $\mathbb{R}$ containing the coordinates of all the points $P_0, P_1, \ldots, P_k$ so that these points are all $\mathbb{F}$-points (from Definition 6.3.1).

If we perform an operation of type (i) or (ii) of 5.3.1, we might obtain several new points $P_{k+1}, \ldots, P_{k+t}$. The lines and circles which appear in (i) and (ii) are $\mathbb{F}$-lines and $\mathbb{F}$-circles, and so each of the new points obtained is at the intersection of

(1) two $\mathbb{F}$-lines,

or (2) an $\mathbb{F}$-line and an $\mathbb{F}$-circle,

or (3) two $\mathbb{F}$-circles.

The next lemma tells us that, at worst, the coordinates of these new points involve extra square roots. ∎

6.3.5 Lemma.
Let $\mathbb{F}$ be a subfield of $\mathbb{R}$.

(a) If two $\mathbb{F}$-lines intersect in a single point, then the point is an $\mathbb{F}$-point.

(b) Given an $\mathbb{F}$-line and an $\mathbb{F}$-circle, there exists a positive number $\alpha \in \mathbb{F}$ such that the points of intersection (if any) of this line and circle are $\mathbb{F}(\sqrt{\alpha})$-points.

(c) Given two $\mathbb{F}$-circles, there exists a positive number $\alpha \in \mathbb{F}$ such that any points of intersection of these two circles are $\mathbb{F}(\sqrt{\alpha})$-points.

Proof. Note that $\mathbb{F}$-lines have equations of the form

$$dx + ey + f = 0 \qquad (1)$$

for d, e, f in $\mathbb{F}$, while $\mathbb{F}$-circles have equations of the form

$$x^2 + y^2 + px + qy + r = 0 \qquad (2)$$

for p, q, r in $\mathbb{F}$.

(a) To find where two $\mathbb{F}$-lines meet, we solve two simultaneous equations of the form (1). This can be done just using field operations $+, -, ., /$, so the solutions are both in $\mathbb{F}$.

(b) Finding where a line and a circle meet amounts to solving two equations, one of the form (1) and the other like (2). This is easily done by using the
$$\frac{-b \pm \sqrt{b^2 - 4ac}}{2a}$$
formula for solving the quadratic $ax^2 + bx + c = 0$. Since, at worst, square roots are introduced, we obtain $\mathsf{F}(\sqrt{\alpha})$-points for some positive $\alpha \in \mathsf{F}$.

(c) The two F-circles
$$x^2 + y^2 + p_1 x + q_1 y + r_1 = 0,$$
$$x^2 + y^2 + p_2 x + q_2 y + r_2 = 0$$
meet where the first of these and the F-line
$$(p_1 - p_2)x + (q_1 - q_2)y + (r_1 - r_2) = 0$$
meet. So this case follows from (b). ∎

With these preliminaries out of the way, we can now prove the "All Constructibles Come From Square Roots" Theorem.

Proof of Theorem 6.1.1. Assume that γ is constructible so that, by the definition of constructible number (Definition 5.4.1), there exists a set of points
$$P_0, P_1, P_2, \ldots, P_n,$$
constructed as in Section 5.3, with $|\gamma|$ equal to the distance between two of these points. Because the initial points P_0 and P_1 are at a distance 1 unit apart, we can introduce coordinate axes so that $P_0 = (0,0)$ and $P_1 = (1,0)$. Note that the coordinates of P_0 and P_1 are in $\mathbb{Q}$.

> Initially, the points P_0, P_1 are $\mathbb{Q}$-points. Put $\mathsf{F}_1 = \mathbb{Q}$. Assume that for some k satisfying $1 \leq k \leq n-1$, the points $P_0, P_1, \ldots, P_k$ are F_k-points, where F_k is a subfield of $\mathbb{R}$.

It follows from the Construction Rules in 5.3.1 that the next point, P_{k+1}, will be at the intersection of

	(1)	a pair of $\mathbb{F}_k$-lines,
or	(2)	an $\mathbb{F}_k$-line and an $\mathbb{F}_k$-circle,
or	(3)	a pair of $\mathbb{F}_k$-circles.

Hence, by Lemma 6.3.5, P_{k+1} is an $\mathbb{F}_{k+1}$-point where

$$\boxed{\mathbb{F}_{k+1} = \mathbb{F}_k(\sqrt{\gamma_k}), \quad \text{for some positive } \gamma_k \text{ in } \mathbb{F}_k.}$$

Since $\mathbb{F}_k \subseteq \mathbb{F}_{k+1}$, this implies further that

$$\boxed{P_0, P_1, \ldots, P_k, P_{k+1} \text{ are } \mathbb{F}_{k+1}\text{-points.}}$$

From the boxed statements, it now follows by mathematical induction on k that the numbers $\gamma_1, \ldots, \gamma_n$ are as described in Theorem 6.1.1. It also follows that $|\gamma|$, being the distance between two of the points

$$P_0, P_1, \ldots, P_n$$

(which are $\mathbb{F}_n$-points), will be the square root of an element of $\mathbb{F}_n$. Thus, for some $\gamma_n \in \mathbb{F}_n$ with $\gamma_n > 0$,

$$\gamma \in \mathbb{F}_n(\sqrt{\gamma_n}). \qquad \blacksquare$$

——————————— **Exercises 6.3** ———————————

1. [This illustrates case (a) of the proof of Lemma 6.3.5.]
   Show that the point where the two $\mathbb{Q}(\sqrt{2})$-lines

   $$\left(1+\sqrt{2}\right)x + y + 3 - \sqrt{2} = 0,$$
   $$2\sqrt{2}x + \left(4+\sqrt{2}\right)y + 1 = 0$$

   meet is a $\mathbb{Q}(\sqrt{2})$-point.

2. [This is the general proof of case (a) of the proof of Lemma 6.3.5.]
   Consider two F-lines
   $$d_1 x + e_1 y + f_1 = 0 \qquad (1)$$
   $$d_2 x + e_2 y + f_2 = 0 \qquad (2)$$
   (with $d_1, d_2, e_1, e_2, f_1, f_2 \in \mathbb{F}$) which are not parallel.
   (i) If $e_1 \neq 0$, use (1) to get an expression for y, substitute this into (2) and hence find where the two lines meet. Explain why the intersection point is an F-point.
   (ii) If $e_1 = 0$, show that the lines meet at an F-point.
   (iii) In (i), you needed to divide by $d_1 e_2 - d_2 e_1$ or something similar. Why can you do this?

3. [This illustrates case (b) of the proof of Lemma 6.3.5.]
   Find where the Q-circle
   $$x^2 + y^2 - 4x - 2y - 4 = 0$$
   and the Q-line
   $$x - y + 3 = 0$$
   meet. For which α is this point of intersection a $\mathbb{Q}(\sqrt{\alpha})$-point?

4. Fill in the details of a general proof of case (b) of the proof of Lemma 6.3.5.

Additional Reading for Chapter 6

Many algebra textbooks contain a proof of the impossibility of the three geometrical constructions. You may like to compare our treatment with those in [BB], [AC], [JF], [WG], [CH] and [LS]. A self-contained account of Problem III is given in [EH].

The algebraic machinery developed in this book can also be used to characterize those regular polygons which can be constructed using straightedge and compass. See, for example, Section 48.2 of [JF], Sections 135-138 of [AC] and Sections 2.5 and 2.7 of [CH].

[BB] B. Bold, *Famous Problems of Geometry and How to Solve Them*, Dover, New York, 1969.

[AC] A. Clark, *Elements of Abstract Algebra*, Wadsworth, Belmont, California, 1971.

[JF] J.B. Fraleigh, *A First Course in Abstract Algebra*, 3rd edition, Addison-Wesley, Reading, Massachusetts, 1982.

[WG] W.J. Gilbert, *Modern Algebra with Applications*, Wiley, New York, 1976.

[CH] C.R. Hadlock, *Field theory and its Classical Problems*, Carus Mathematical Monographs, No. 19, Mathematical Association of America, 1978.

[EH] E.W. Hobson, *Squaring the Circle*, Cambridge University Press, 1913; reprinted in *Squaring the Circle, and Other Monographs*, Chelsea, 1953.

[WL] W. Ledermann, *Complex Numbers*, Library of Mathematics, Routledge and Kegan Paul, London, 1965.

[LS] L.W. Shapiro, *Introduction to Abstract Algebra*, McGraw-Hill, New York, 1975.

CHAPTER 7

Transcendence of e and π

The main purpose of this chapter is to prove that the number π is transcendental, thereby completing the proof of the impossibility of squaring the circle (Problem III of the Introduction). We first give the proof that e is a transcendental number, which is somewhat easier. This is of considerable interest in its own right, and its proof introduces many of the ideas which will be used in the proof for π. With the aid of some more algebra – the theory of symmetric polynomials – we can then modify the proof for e to give the proof for π.

The proof for e uses elementary facts about integration. (We expect you are familiar with these facts from a course on calculus.) The proof for π uses integration of complex-valued functions, which we introduce in this chapter.

7.1 Preliminaries

This section contains the preliminary results which are needed for our proofs that e and π are transcendental numbers. The extra preliminaries which are used only in the proof for π will be given later.

Factors and Prime Numbers

Let a and b be integers. Recall that b is said to be a *factor* of a if there is an integer c such that $a = bc$.

The integer 0 is exceptional in that it has every integer as a factor (if $a = 0$ then every b satisfies $a = bc$ for $c = 0$). Another way to say this is that

> *if a is an integer which fails to have some integer b as a factor, then $a \neq 0$.*

This statement provides a way of proving that an integer is not zero, and it will play an important rôle in our proof. This leads us to consider ways in which we can show that a given integer fails to have some other given integer as a factor.

One consequence of the definition of a factor is that if a is nonzero and has b as a factor then $|a| \geq |b|$. Hence

> *if $b > |a|$ and $a \neq 0$, then b cannot be a factor of a.*

Another consequence of the definition of a factor is that if an integer b is a factor of two integers, then it is also a factor of their sum, their difference and their product. Hence

> *if b is a factor of an integer a_1 but b is not a factor of an integer a_2, then b is not a factor of $a_1 + a_2$*

as otherwise b would be a factor of the difference $(a_1 + a_2) - a_1 = a_2$.

Recall that an integer p is said to be a *prime* if $p > 1$ and if p has no positive factors other than 1 and p itself. The Fundamental Theorem of Arithmetic is stated in Exercises 1.4 #9. It can be shown from this theorem that if a prime number p is a factor of the product ab of two integers a and b, then p must be a factor of a or a factor of b (or a factor of both a and b).

From this it is easy to deduce another useful test for failure to be a factor of a product:

if a prime p is not a factor of any of the integers $a_1, a_2, \ldots, a_m$ then it is not a factor of their product $a_1 a_2 \ldots a_m$.

Another fact which we shall need later is given by the following proposition, whose proof was known to Euclid.

7.1.1 Proposition. *There are infinitely many prime numbers.*

Proof. To prove this we note that there is at least one prime number (2 – for example) and we shall prove that given any finite set of prime numbers

$$\{p_1, p_2, \ldots, p_n\} \qquad (1)$$

there is always another prime number which is not in the set. To this end consider the integer

$$K = p_1 p_2 \ldots p_n + 1. \qquad (2)$$

Each p_i is a factor of the first term on the right-hand side of (2), but not of the second term, 1. Hence p_i is not a factor of K. But K does have a prime factor q by the Fundamental Theorem of Arithmetic. Thus there is a prime q which is not in the set (1), as we wished to show. ∎

This means that, given any integer m, no matter how large, we can be sure that there is a prime (indeed, infinitely many primes) larger than m. We use this fact frequently in this chapter.

Calculus

We expect that all of our readers will have met the following properties of integrals in a calculus course. (Proofs can be found in many texts: see for example Theorems 5 and 6 in Chapter 13 of [MS].)

7.1.2 Theorem. *Let f, g and $h_1, \ldots, h_n$ be functions from $\mathbb{R}$ to $\mathbb{R}$ and let $a, b \in \mathbb{R}$.*

(i) [Linearity of Integration] *If $d_1, \ldots, d_n \in \mathbb{R}$ and if each of the following integrals exists, then*

$$\int_a^b \bigl(d_1 h_1(x) + \ldots + d_n h_n(x)\bigr) dx = d_1 \int_a^b h_1(x) dx + \ldots + d_n \int_a^b h_n(x) dx.$$

(ii) [Integration by Parts] *If f and g are functions such that each of the following integrals exists, then*

$$\int_a^b f(x)g'(x)dx = f(b)g(b) - f(a)g(a) - \int_a^b f'(x)g(x)dx.$$ ∎

We shall use these results only in the cases where the functions are polynomial or exponential functions, or products thereof; the integrals of such functions always exist.

The proof of the next result depends on the idea of the degree of a nonzero polynomial form, which was explained in Chapter 1.

7.1.3 Lemma. *Let $g(X)$ and $h(X)$ be polynomials whose coefficients are real numbers. If*

$$g(x) = h(x)e^{-x} \qquad \text{for all } x \in \mathbb{R} \tag{3}$$

then $g(X)$ and $h(X)$ are both equal to the zero polynomial.

Proof. It follows from (3) that $h(x) = g(x)e^x$ and so differentiating both sides gives

$$h'(x) = \Big(g(x) + g'(x)\Big)e^x \qquad \text{for all } x \in \mathbb{R}.$$

If we multiply both sides by $g(x)$ and then replace the factor $e^x g(x)$ on the right-hand side by $h(x)$, we find that

$$\begin{aligned} h'(x)g(x) &= \Big(g(x) + g'(x)\Big)e^x g(x) \\ &= g(x)h(x) + g'(x)h(x) \qquad \text{for all } x \in \mathbb{R}. \end{aligned}$$

Rearranging this equation and interpreting it in terms of polynomial forms gives

$$g(X)h(X) = h'(X)g(X) - g'(X)h(X). \tag{4}$$

If either of the polynomials $g(X)$ or $h(X)$ is zero, then so is the other by (3); hence it only remains to prove the lemma in the case where both $g(X)$ and $h(X)$ are nonzero polynomials.

Now the degree of $g'(X)$ is either undefined or is one less than the degree of $g(X)$, and similarly the degree of $h'(X)$ is either undefined or is one less than the degree of $h(X)$. Hence the equation (4) gives a contradiction because the degree of its right side either is undefined or is less than the degree of $g(X)h(X)$. Thus the case in which both of the polynomials $g(X)$ and $h(X)$ are nonzero cannot occur. ∎

Transcendence

The following lemma concerning the limit of a sequence will also be needed.

7.1.4 Lemma. *For each real number $c \geq 0$, $\lim_{n \to \infty} \dfrac{c^n}{n!} = 0$.*

Proof. Let m be an integer such that $m > 1$ and $m \geq 2c$. For all integers $n \geq m$,

$$\frac{c^n}{n!} = \frac{c}{1}\frac{c}{2}\cdots\frac{c}{m-1}\frac{c}{m}\frac{c}{m+1}\cdots\frac{c}{n}$$
$$\leq \frac{c}{1}\frac{c}{2}\cdots\frac{c}{m-1}\frac{1}{2}\frac{1}{2}\cdots\frac{1}{2} = \frac{d}{2^{n-m+1}}$$

where d is the constant $\dfrac{c}{1}\dfrac{c}{2}\cdots\dfrac{c}{m-1}$.

Since $\dfrac{d}{2^{n-m+1}}$ approaches 0 as n approaches ∞, so does $\dfrac{c^n}{n!}$. ∎

Expanding a polynomial in powers of $(X - r)$

Let $g(X)$ be a polynomial over a field so that

$$g(X) = c_0 + c_1 X + c_2 X^2 + \ldots + c_m X^m \tag{5}$$

for some integer $m \geq 0$ and coefficients $c_0, c_1, c_2, \ldots, c_m$ in the field. It is often very useful to know that, given r in the field, we can express $g(X)$ in the form

$$g(X) = b_0 + b_1(X - r) + b_2(X - r)^2 + \ldots + b_m(X - r)^m \tag{6}$$

where the coefficients $b_0, b_1, b_2, \ldots, b_m$ are also in the field.

Whereas (5) expresses the polynomial as a linear combination of powers of X, (6) expresses it as a linear combination of the powers of $(X - r)$ and is called the *expansion of $g(X)$ in powers of $(X - r)$*.

A general procedure for expanding a polynomial in powers of $(X - r)$ is to expand each monomial X^i in powers of $(X - r)$ using the Binomial Theorem, and then collect like terms. We illustrate this procedure in the next example.

Note that the coefficients b_i in (6) turn out to be polynomials $d_i(X)$ evaluated at r; that is, $b_i = d_i(r)$.

7.1.5 Example. To expand the polynomial (over $\mathbb{R}$)
$$g(X) = 3 + 4X + 7X^2 + X^3$$
in powers of $(X-r)$, we write
$$\begin{aligned}
g(X) &= g(X - r + r) \\
&= 3 + 4[(X - r) + r] + 7[(X - r) + r]^2 + [(X - r) + r]^3 \\
&= 3 + 4[(X - r) + r] + 7[(X - r)^2 + 2r(X - r) + r^2] + \\
&\quad [(X - r)^3 + 3r(X - r)^2 + 3r^2(X - r) + r^3] \\
&= (3 + 4r + 7r^2 + r^3) + (4 + 14r + 3r^2)(X - r) + \\
&\quad (7 + 3r)(X - r)^2 + (X - r)^3.
\end{aligned}$$

Thus we have shown that the polynomial $g(X)$ can be expanded in powers of $(X-r)$ to give
$$g(X) = d_0(r) + d_1(r)(X - r) + d_2(r)(X - r)^2 + d_3(r)(X - r)^3$$
where $d_0(r), d_1(r), d_2(r)$, and $d_3(r)$ are the values at r of polynomials $d_0(X),\ldots,d_3(X)$ given by
$$\begin{aligned}
d_0(X) &= 3 + 4X + 7X^2 + X^3, \\
d_1(X) &= 4 + 14X + 3X^2, \\
d_2(X) &= 7 + 3X, \\
d_3(X) &= 1.
\end{aligned}$$

In particular we have, as in (6),
$$g(X) = b_0 + b_1(X - r) + b_2(X - r)^2 + b_3(X - r)^3$$
where $b_0 = d_0(r)$, $b_1 = d_1(r)$, $b_2 = d_2(r)$, and $b_3 = d_3(r)$. ■

The general result, whose proof is obvious from the above example, is as follows.

7.1.6 Proposition. *If $g(X)$ is a polynomial of degree at most n, whose coefficients are integers, then there exist polynomials*
$$d_0(X),\ldots,d_n(X)$$
with integer coefficients and degree at most n such that, for all $r \in \mathbb{R}$,
$$g(X) = d_0(r) + d_1(r)(X - r) + \ldots + d_n(r)(X - r)^n.$$ ■

Note that the same polynomials $d_0(X),\ldots,d_n(X)$ work for all values of r.

Transcendence

The following result confirms that when two polynomials in $(X-r)$ are equal, we can equate coefficients (just as we can for ordinary polynomials, which is the case $r = 0$).

7.1.7 Lemma. Let $r \in \mathbb{R}$ and $b_i, c_i \in \mathbb{R}$ for $i \in \{0, 1, \ldots, n\}$. If

$$c_0 + c_1(X - r) + \ldots + c_n(X - r)^n = b_0 + b_1(X - r) + \ldots + b_n(X - r)^n,$$

then

$$c_0 = b_0, \; c_1 = b_1, \; \ldots, \; c_n = b_n.$$

Proof. We have

$$c_0 + c_1(x - r) + \ldots + c_n(x - r)^n = b_0 + b_1(x - r) + \ldots + b_n(x - r)^n$$

for all $x \in \mathbb{R}$. Substituting $x = r$ gives $c_0 = b_0$. Differentiating with respect to x and substituting $x = r$ then gives $c_1 = b_1$. Repeating this a further $n - 1$ times gives the required result. ∎

Approximating Real Numbers by Rational Numbers

That we can approximate any real number a by rational numbers is well-known. We say that the set of rational numbers $\mathbb{Q}$ is a *dense* subset of the real line $\mathbb{R}$; that is, given any interval containing a, there is a rational number which lies in that interval. More formally this means that, given any real number a and a "small" number $\varepsilon > 0$, we can find integers M and M_1 and a number ε_1 such that $|\varepsilon_1| < \varepsilon$ and

$$a = \frac{M_1}{M} + \varepsilon_1. \tag{7}$$

This says that there is a rational number M_1/M at a distance of $|\varepsilon_1| < \varepsilon$ from the given real number a.

For our purposes, a more relevant type of approximation, is suggested by the following question: *Given a "small" number ε, can we find integers M and M_1 and a nonzero number ε_1 with $|\varepsilon_1| < \varepsilon$ such that*

$$a = \frac{M_1 + \varepsilon_1}{M}? \tag{8}$$

Thus we are asking not only that the distance from a to M_1/M be small, but also that it be small even when multiplied by M, the denominator of the approximating fraction.

7.1.8 Example. It is easy to see that the number e can be approximated as in (8). To see this first note that the well-known Taylor series for the exponential function gives
$$e = 1 + \frac{1}{1!} + \frac{1}{2!} + \ldots + \frac{1}{n!} + \frac{1}{(n+1)!} + \ldots .$$
For each integer $n \geq 1$, let
$$A_n = 1 + \frac{1}{1!} + \frac{1}{2!} + \ldots + \frac{1}{n!}.$$
Hence
$$e = A_n + \frac{1}{(n+1)!}\left[1 + \frac{1}{n+2} + \frac{1}{(n+2)(n+3)} + \ldots\right]$$
$$\leq A_n + \frac{1}{(n+1)!}\left[1 + \frac{1}{2} + \frac{1}{2^2} + \ldots\right]$$
$$= A_n + \frac{2}{(n+1)!}.$$

Given $\varepsilon > 0$ choose n to be any integer $> 2/\varepsilon$, so that $2/n < \varepsilon$. Put $M = n!$ and $M_1 = (n!)A_n$. You should now be able to see how to choose ε_1 so that (8) holds. We leave the details as Exercises 7.1 #8.■

Can all real numbers a be approximated as in (8) by rationals? That the answer to this question is NO is clear from the following proposition.

7.1.9 Proposition. *Let a be a real number with the following property: for each real number $\varepsilon > 0$ there are integers M and M_1, and a nonzero real number ε_1 with $|\varepsilon_1| < \varepsilon$ such that*
$$a = \frac{M_1 + \varepsilon_1}{M}.$$
Then the real number a is irrational.

Proof. Suppose, contrary to the conclusion of the proposition, that a is a rational number, say $a = p/q$ where p and q are integers and $q > 0$.

Let $\varepsilon = 1/(2q)$. By the hypothesis of the proposition there exist integers M and M_1, and a nonzero real number ε_1 with $|\varepsilon_1| < \varepsilon$ such that
$$\frac{p}{q} = \frac{M_1 + \varepsilon_1}{M}, \quad \text{and hence} \quad pM - qM_1 = q\varepsilon_1.$$
The left hand side of the last equation is an integer, while $0 < |q\varepsilon_1| < \frac{1}{2}$. This contradiction shows that a cannot be rational. ■

From Example 7.1.8 and Proposition 7.1.9 it follows that e is an irrational number. Note also that requiring ε_1 to be nonzero is crucial in the proof of the above proposition. By requiring ε_1 to be nonzero, we are not allowing the rational approximation M_1/M to equal the number a.

A very useful idea which this proposition suggests is that *the "better" a real number can be approximated by rational numbers (not equal to it), the "worse" that real number must be.* Here we are thinking of the rational numbers as "good", the irrational numbers which are algebraic as "bad", and the irrational numbers which are not algebraic as "worse". We shall follow up this idea in the next section.

──────────── Exercises 7.1 ────────────

1. Let n be a positive integer. Verify that each of the following numbers has $(n-1)!$ as a factor.
   (i) $n!$.
   (ii) $n! + (n+1)! + \ldots + (n+m)!$ where m is a positive integer.
   (iii) $(2n)!$.

2. Prove, from the definition of factor, that if an integer b is a factor of another integer a which is not zero, then $|a| \geq |b|$.

3. Prove, from the definition of factor, that if an integer b is a factor of each of the integers a_1 and a_2 then b is also a factor of the integers $a_1 + a_2$, $a_1 - a_2$, and $a_1 a_2$. Is b always a factor of a_1/a_2?

4. Let m and n be integers with $m > n$. In each of the following cases state whether the integer always has m as a factor. (Justify your answers.)
   (i) $mn - n$.
   (ii) $nm - m$.

5. Let p be a prime number and let m be a positive integer such that $m < p$. Show that p is not a factor of
   (i) $m(m-1)$;
   (ii) $m!$;
   (iii) $(m!)^p$.

6. Let $f(x) = a_0 + a_1 x + \ldots + a_n x^n$ be a polynomial function with real coefficients. The result of differentiating this polynomial i times in succession is denoted by $f^{(i)}$. It is called the i-th derivative of the polynomial f. Show that $f^{(n+1)}(x) = 0$ for all $x \in \mathbb{R}$.

7. (a) Express $f(X) = 2 - 4X + 3X^2 - 7X^3$ as a polynomial in $(X - r)$. Write down the formulae for the polynomials $d_0(X), \ldots, d_3(X)$ in Proposition 7.1.6.

   (b) Repeat (a) with $f(X) = c_0 + c_1 X + c_2 X^2 + c_3 X^3 + c_4 X^4$ where $c_0, \ldots, c_4 \in \mathbb{Z}$.

8. Complete the details of Example 7.1.8. Why is M_1 an integer?

9. (a) Show that the rational number $3/7$ can be approximated as in (7) of Section 7.1.

   [Hint. What can M_1 and M be if $\varepsilon = 1/100$?]

   (b) Show that every rational number can be approximated as in (7) of Section 7.1.

10. Would Proposition 7.1.9 be true if the word "nonzero" were omitted? (Justify your answer.)

11. Prove that each of the numbers $\cos(1)$ and $\sin(1)$ is irrational by using the Taylor series for the functions cos and sin.

7.2 e is Transcendental

An important property of the number e is that, if $g(x) = e^x$ for all $x \in \mathbb{R}$, then $g'(x) = g(x)$ for all x. Indeed this fact distinguishes the number e from all other positive real numbers (as shown in Exercises 7.2 #1 below). We shall also use the fact that $e^{x+y} = e^x e^y$ for all real numbers x and y (a property of e which is shared by all positive reals).

Our proof that the number e is transcendental (see Definition 2.1.5) shall begin by supposing, on the contrary, that e is algebraic over $\mathbb{Q}$. We let

$$t(X) = a_0 + a_1 X + \ldots + a_{m-1} X^{m-1} + X^m$$

denote irr(e,$\mathbb{Q}$), the irreducible polynomial of e over $\mathbb{Q}$. Then we have $m \geq 1$ and $a_i \in \mathbb{Q}$ for each i, and

$$0 = t(\mathrm{e}) = a_0 + a_1\mathrm{e} + \ldots + a_{m-1}\mathrm{e}^{m-1} + \mathrm{e}^m. \qquad (*)$$

Now clearly $a_0 \neq 0$ as otherwise X would be a factor of the polynomial $t(X)$, which is irreducible (and which is not just X since $\mathrm{e} \neq 0$).

If we now multiply $(*)$ by the product of the denominators of the rational numbers $a_0, a_1, \ldots, a_{m-1}$ we get

$$c_0 + c_1\mathrm{e} + \ldots + c_m\mathrm{e}^m = 0 \qquad (1)$$

where $c_0, c_1, \ldots, c_m$ are integers such that $c_0 \neq 0$ and $c_m \neq 0$.

In due course, we shall derive a contradiction from this.

Idea of Proof: M's and c's

To show that (1) leads to a contradiction, we shall use the idea (mentioned in Section 7.1) that the "worse" a number is, the "better" it can be approximated by rationals. As we have seen in Example 7.1.8 and Proposition 7.1.9, the irrationality of e follows very simply from the following property: *For each $\varepsilon > 0$ there is a nonzero number ε_1 with $|\varepsilon_1| < \varepsilon$ and integers M and M_1 such that*

$$\mathrm{e} = \frac{M_1 + \varepsilon_1}{M}.$$

The aim now is to extend this "close approximation" by rationals to each of the remaining powers of e in the equation (1). The aim is thus:

Given $\varepsilon > 0$, to show there are numbers $\varepsilon_1, \ldots, \varepsilon_m$ all less than ε in absolute value, and integers $M, M_1, \ldots, M_m$ such that

$$\mathrm{e} = \frac{M_1 + \varepsilon_1}{M}, \quad \mathrm{e}^2 = \frac{M_2 + \varepsilon_2}{M}, \quad \ldots \quad, \mathrm{e}^m = \frac{M_m + \varepsilon_m}{M}.$$

Substituting these powers of e into (1) and rearranging the terms gives

$$[c_0 M + c_1 M_1 + \ldots + c_m M_m] + [c_1 \varepsilon_1 + \ldots + c_m \varepsilon_m] = 0. \qquad (2)$$

Because M is an integer and so is each c_i and each M_i, the first sum on the left side of (2) is an integer. If we can show this integer is nonzero we shall get the desired contradiction: we can choose each ε_i so small that the second sum on the left side of (2) has its absolute value less than $\frac{1}{2}$ and hence is unable to cancel out the first sum.

It now remains to show how to find suitable numbers $\varepsilon_1, \ldots, \varepsilon_m$ and $M, M_1, \ldots, M_m$.

Producing the ε's and the M's from integrals

The following lemma shows how an integer (namely $k!$) arises from an integral involving the exponential function. This lemma therefore suggests the possibility of using integrals to construct the required integers $M, M_1, \ldots, M_m$.

7.2.1 Lemma. *For each integer $k \geq 0$ there are polynomials $g_k(X)$ and $h_k(X)$ with real coefficients such that, for all $r \in \mathbb{R}$,*

(a) $$\int_0^r x^k e^{-x} dx = k! - e^{-r} g_k(r),$$
(b) $$\int_0^r (x-r)^k e^{-x} dx = h_k(r) - e^{-r} k!.$$

Proof. This can be proved by mathematical induction on k, using integration by parts (see Theorem 7.1.2). ∎

Motivated by the above lemma, we shall aim at getting the integers M and $M_1, \ldots, M_m$ from integrals of the form

$$\int_0^r f(x) e^{-x} dx \tag{3}$$

for $r \in \{1, 2, \ldots, m\}$, where f is a polynomial function. In the following example, we show how to apply the above lemma to such integrals.

7.2.2 Example. Let $f(X) = 3 + 4X - 10X^3$. By linearity of integration (Theorem 7.1.2), and then Lemma 7.2.1(a), for all $r \in \mathbb{R}$,

$$\int_0^r f(x) e^{-x} dx$$
$$= \int_0^r (3 + 4x - 10x^3) e^{-x} dx$$
$$= 3 \int_0^r 1 \cdot e^{-x} dx + 4 \int_0^r x e^{-x} dx - 10 \int_0^r x^3 e^{-x} dx$$
$$= 3(0! - e^{-r} g_0(r)) + 4(1! - e^{-r} g_1(r)) - 10(3! - e^{-r} g_3(r))$$
$$= (3 \cdot 0! + 4 \cdot 1! - 10 \cdot 3!) - e^{-r}(3g_0(r) + 4g_1(r) - 10g_3(r))$$

where $g_0(X), g_1(X)$ and $g_3(X)$ are polynomials. Hence

$$\int_0^r f(x) e^{-x} dx = M - e^{-r} G(r)$$

where $M = 3 \cdot 0! + 4 \cdot 1! - 10 \cdot 3! = -53$ and $G(X)$ is the polynomial

$$G(X) = 3g_0(X) + 4g_1(X) - 10g_3(X).$$ ∎

The next lemma generalizes this example.

7.2.3 Lemma. *Given a polynomial $f(X)$ with real coefficients, there is a unique real number M and a unique polynomial $G(X)$ with real coefficients such that, for all $r \in \mathbb{R}$,*

$$\int_0^r f(x)e^{-x}dx = M - e^{-r}G(r).$$

Indeed M and $G(X)$ are unique in the strong sense that, if $P_1(X)$ and $P_2(X)$ are polynomials with real coefficients such that

$$\int_0^r f(x)e^{-x}dx = P_1(r) - e^{-r}P_2(r)$$

for all $r \in \mathbb{R}$, then $P_2(X) = G(X)$ and $P_1(X) = M$ (so that $P_1(X)$ is a constant polynomial).

Proof. The existence of M and $G(X)$ follows as in Example 7.2.2 by expanding $f(X)$ in powers of X, using linearity of integration (Theorem 7.1.2(i)) and then applying Lemma 7.2.1(a).

To prove uniqueness, we assume that the two formulae in the statement of the lemma hold. Then, for all $r \in \mathbb{R}$,

$$M - e^{-r}G(r) = P_1(r) - e^{-r}P_2(r)$$

and hence $M - P_1(r) = e^{-r}\big(G(r) - P_2(r)\big)$. It follows from Lemma 7.1.3 that $M - P_1(X)$ and $G(X) - P_2(X)$ must both be the zero polynomial, which gives $P_1(X) = M$ and $P_2(X) = G(X)$, as required. ∎

We now use this lemma to define the numbers we need for the transcendence proof.

7.2.4 Definition. In Lemma 7.2.3, choose f to be the polynomial

$$f(X) = \frac{X^{p-1}}{(p-1)!}(X-1)^p(X-2)^p\ldots(X-m)^p \qquad (4)$$

where p is a prime number (which we shall later take to be rather large) and m is the integer occurring in (1). Further, choose

(i) M and $G(X)$ to be the number and polynomial given by Lemma 7.2.3,

(ii) $M_r = G(r)$ for $r \in \{1, 2, \ldots, m\}$, and

(iii) $\varepsilon_r = e^r \int_0^r f(x)e^{-x}dx$ for $r \in \{1, 2, \ldots, m\}$. ∎

Note that, from (4), the degree n of $f(X)$ is
$$n = mp + p - 1. \tag{5}$$
Also, by Proposition 7.1.6, there are polynomials $d_0, \ldots, d_n$ with integer coefficients and degree at most n such that, for all $r \in \mathbb{R}$,
$$f(X) = \frac{1}{(p-1)!}\Big(d_0(r) + d_1(r)(X-r) + \ldots + d_n(r)(X-r)^n\Big). \tag{6}$$

7.2.5 Lemma. With the notation as in (6), if $r \in \{1, 2, \ldots, m\}$ then $d_0(r) = d_1(r) = \ldots = d_{p-1}(r) = 0$.

Proof. Fix $r \in \{1, 2, \ldots, m\}$. Because $(X-r)^p$ is a factor of $f(X)$ by (4), we have
$$f(X) = \frac{1}{(p-1)!}(X-r)^p g(X)$$
for some polynomial $g(X)$ with integer coefficients. If we expand $g(X)$ in powers of $(X-r)$ (see Proposition 7.1.6) and multiply each resulting term by $(X-r)^p$ we see that
$$f(X) = \frac{1}{(p-1)!}\Big(b_p(X-r)^p + b_{p+1}(X-r)^{p+1} + \ldots + b_n(X-r)^n\Big) \tag{7}$$
for some real numbers $b_p, \ldots, b_n$. By Lemma 7.1.7 we can equate the coefficients of $1, (X-r), \ldots, (X-r)^n$ in (7) and (6). Since the coefficients of $1, (X-r), \ldots, (X-r)^{p-1}$ are zero in (7), they must be zero in (6) also. ∎

Properties of $M, M_1, \ldots, M_m$ and $\varepsilon_1, \ldots, \varepsilon_m$

We now derive properties of the numbers which were introduced in Definition 7.2.4.

7.2.6 Lemma. (a) The number M is an integer which does not have the prime p as a factor when $p > m$.

(b) There is a polynomial G_1 with integer coefficients and degree at most n such that $G(r) = p\,G_1(r)$ for $r \in \{1, 2, \ldots, m\}$. In particular, $M_1, \ldots, M_m$ are all integers with the prime p as a factor.

Proof. (a) The polynomial $f(X)$ defined in (4) can be expanded in powers of X so that we can write
$$f(X) = \frac{1}{(p-1)!}\Big(a_{p-1}X^{p-1} + a_p X^p + \ldots + a_n X^n\Big) \tag{8}$$

Transcendence

where $a_{p-1}, \ldots, a_n$ are integers with $a_{p-1} = \pm(m!)^p \neq 0$.

By the linearity of integration (Theorem 7.1.2(i)), it follows from (8) that

$$\int_0^r f(x)e^{-x}dx$$
$$= \int_0^r \frac{1}{(p-1)!}\left(a_{p-1}x^{p-1} + a_p x^p + \ldots + a_n x^n\right)e^{-x}dx$$
$$= \frac{1}{(p-1)!}\left(a_{p-1}\int_0^r x^{p-1}e^{-x}dx + a_p \int_0^r x^p e^{-x}dx + \ldots + a_n \int_0^r x^n e^{-x}dx\right).$$

By Lemma 7.2.1(a), each of the integrals in the above sum is the difference of two terms, the second of which involves e^{-r}. Hence, by the method used in Example 7.2.2, $\int_0^r f(x)e^{-x}dx$ is the difference of two sums, the first of which is

$$\frac{1}{(p-1)!}\left(a_{p-1}(p-1)! + a_p p! + \ldots + a_n n!\right)$$

while the second is of the form $e^{-r} \times$ *polynomial in* r. By (i) of Definition 7.2.4 and the uniqueness part of Lemma 7.2.3, M is this first term. Hence, dividing $(p-1)!$ into each term in the above sum gives

$$M = a_{p-1} + a_p p + \ldots + a_n n(n-1)\ldots(p+1)p.$$

Since the coefficients $a_{p-1}, a_p, \ldots, a_n$ are all integers, the number M is an integer. Also the prime p is a factor of every term in the above sum except the first term $a_{p-1} = \pm(m!)^p$, which cannot contain p as a factor when $p > m$ (by Exercises 7.1 #5); hence p cannot be a factor of M when $p > m$.

(b) [Here we use the expansion of $f(X)$ in powers of $X - r$, as given in (6) above.] By the linearity of integration, it follows from (6) that

$$\int_0^r f(x)e^{-x}dx = \frac{1}{(p-1)!}\Big(d_0(r)\int_0^r e^{-x}dx + d_1(r)\int_0^r (x-r)e^{-x}dx +$$
$$\ldots + d_n(r)\int_0^r (x-r)^n e^{-x}dx\Big).$$

By Lemma 7.2.1(b), each of the integrals in the above sum is the difference of two terms, the second of which involves e^{-r}. Hence $\int_0^r f(x)e^{-x}dx$ is the difference of two sums. The second of these is

$$\frac{e^{-r}}{(p-1)!}\left(d_0(r) + d_1(r)1! + \ldots + d_n(r)n!\right) \qquad (9)$$

and the first is a polynomial in r with real coefficients. By the uniqueness part of Lemma 7.2.3, $G(r)$ is equal to the polynomial part of (9); that is,

$$G(r) = \frac{1}{(p-1)!}\Big(d_0(r) + d_1(r)1! + \ldots + d_n(r)n!\Big).$$

In the special cases where $r \in \{1, 2, \ldots, m\}$, we know from Lemma 7.2.5 that $d_0(r) = d_1(r) = \ldots = d_{p-1}(r) = 0$. Hence

$$G(r) = \frac{1}{(p-1)!}\Big(d_p(r)p! + \ldots + d_n(r)n!\Big).$$

If we divide $(p-1)!$ into each term we see that $G(r) = p\,G_1(r)$ where

$$G_1(X) = d_p(X) + d_{p+1}(X)(p+1) + \ldots + d_n(X)n(n-1)\ldots(p+1).$$

Notice that $G_1(X)$ is a polynomial with integer coefficients and degree at most n since each $d_i(X)$ is such a polynomial. Hence $G_1(r)$ is an integer (since r is) and so, by (ii) of Definition 7.2.4, $M_r = p\,G_1(r)$ is an integer having p as a factor. ∎

The next lemma shows that the ε's defined in Definition 7.2.4 can be made arbitrarily small by choosing p large enough.

7.2.7   Lemma.   *If $f(X)$ is given as a function of p by (4), and $r \in \{1, 2, \ldots, m\}$, then*

$$\lim_{p \to \infty} \int_0^r f(x)e^{-x}dx = 0.$$

Proof.   Let $r \in \{1, 2, \ldots, m\}$ and let $x \in [0, r]$. Using (4) gives

$$\begin{aligned}
|f(x)e^{-x}| &\leq |f(x)| \\
&= \frac{x^{p-1}}{(p-1)!}\Big|(x-1)^p(x-2)^p \ldots (x-m)^p\Big| \\
&\leq \frac{x^{p-1}}{(p-1)!}(x+1)^p(x+2)^p \ldots (x+m)^p \\
&\leq \frac{r^{p-1}}{(p-1)!}(r+1)^p(r+2)^p \ldots (r+m)^p \\
&= \frac{1}{r}\frac{C^p}{(p-1)!},
\end{aligned}$$

Transcendence

where $C = r(r+1)(r+2)\ldots(r+m)$ is independent of p. But

$$\left|\int_0^r f(x)e^{-x}dx\right| \leq (\text{length of interval}) \times (\text{upper bound for } |f(x)e^{-x}|)$$
$$= r \times \frac{1}{r}\frac{C^p}{(p-1)!}.$$

Hence the conclusion of the lemma follows from Lemma 7.1.4. ∎

With the aid of the above lemmas, it is now an easy task to prove that e is transcendental.

7.2.8 Theorem. *e is a transcendental number.*

Proof. Suppose on the contrary that e is algebraic. Then, as shown at the start of this section (see (1) at the beginning of this section), there is an integer $m \geq 1$ and integers $c_0, c_1, c_2, \ldots, c_m$ such that $c_0 \neq 0$, $c_m \neq 0$ and

$$c_0 + c_1 e + c_2 e^2 + \ldots + c_m e^m = 0. \tag{10}$$

Let M and M_r, ε_r, for $r \in \{1, 2, \ldots, m\}$, be as in Definition 7.2.4. From parts (iii), (i) and (ii) of that definition respectively, it follows that

$$e^{-r}\varepsilon_r = \int_0^r f(x)e^{-x}dx = M - e^{-r}G(r) = M - e^{-r}M_r$$

and hence

$$e^r = \frac{M_r + \varepsilon_r}{M}.$$

Substituting these results into (10) gives

$$[c_0 M + c_1 M_1 + \ldots + c_m M_m] + [c_1 \varepsilon_1 + \ldots + c_m \varepsilon_m] = 0. \tag{11}$$

We now choose the prime number p to be larger than each of m and $|c_0|$. As $p > m$ it follows from Lemma 7.2.6(a) that

the integer M does not have p as a factor.

As $p > |c_0|$ it follows that c_0 does not have p as a factor and hence (as explained at the beginning of Section 7.1) that

$c_0 M$ does not have p as a factor.

But each of $M_1, M_2, \ldots, M_m$ is an integer having p as a factor by Lemma 7.2.6(b). Hence p is not a factor of the sum

$c_0 M + c_1 M_1 + \ldots + c_m M_m$, *which is therefore a <u>nonzero</u> integer.* (12)

On the other hand, if we choose p sufficiently large (which is possible by Proposition 7.1.1), then by Definition 7.2.4(iii) and Lemma 7.2.7 we get

$$|c_1\varepsilon_1 + \ldots + c_m\varepsilon_m| < \frac{1}{2}. \tag{13}$$

But (12) and (13) contradict (11). So our supposition is wrong. Hence e is a transcendental number. ∎

7.2.9 Remark. Now that the proof is complete, you might like to reflect on the reasons for the various features in the definition of $f(X)$ in (4).

(i) The $(p-1)!$ in the denominator means that ε_r can be made arbitrarily small by taking p large enough. (See Lemma 7.2.7.)

(ii) The factor X^{p-1} ensures that, despite the $(p-1)!$ in the denominator of $f(X)$, M is an integer. The exact power $p-1$ (rather than a higher power) means that M is not divisible by p if $p > m$. (See the proof of Lemma 7.2.6(a). See also Exercise 3 below.)

(iii) The factors $(X-1)^p, \ldots, (X-m)^p$ ensure that, despite the $(p-1)!$ in the denominator, each of $M_1, \ldots, M_m$ is an integer. (See the proof of Lemma 7.2.6(b).)

(iv) All of the factors $(X-1)^p, \ldots, (X-m)^p$ are required so that we can deal with all the exponents $1, 2, \ldots, m$ in (10).

---------- **Exercises 7.2** ----------

1. [This exercise proves the claim made at the beginning of Section 7.2 that e is distinguished from all other positive real numbers by the fact that the derivative of e^x is e^x.]

    Let c be a positive real number and assume that $g(x) = c^x$ for all $x \in \mathbb{R}$. If $g'(x) = g(x)$ for all real x, prove that $c = e$.

    [Hint. $c^x = (e^{\ln c})^x = e^{x \ln c}$.]

2. Prove Lemma 7.2.1.

3. Show that Lemma 7.2.6 would be false if we omitted the factor X^{p-1} in the definition of the function $f(X)$ given in (4) of the text.

4. Show that Lemma 7.2.6 would be false if we replaced the factor X^{p-1} by the factor X^p in the definition of $f(X)$ given in (4) of the text.

5. (a) Prove that e^2 is a transcendental number.
    [Hint. Use Theorem 7.2.8 and the definition of algebraic.]
   (b) Prove that $e^2 + 3e + 1$ is a nonzero transcendental number.
   (c) Prove that if $f(X)$ is a nonzero polynomial with rational coefficients, then $f(e)$ is a nonzero transcendental number.

6. Prove that $\sqrt{e}$ is transcendental. [Hint. Use Corollary 4.4.4.]

7. Deduce from Lemma 7.2.1(a) that $\int_0^\infty x^k e^{-x} dx = k!$.
   [Hint. You may assume that $\lim_{r \to \infty} e^{-r} g(r) = 0$ for each polynomial $g(X)$ with real coefficients.]

8. (a) Deduce from Definition 7.2.4, Lemma 7.2.3 and the hint for Exercise 7 that
$$M = \int_0^\infty f(x) e^{-x} dx.$$
   (b) Using (a) deduce from Definition 7.2.4 and Lemma 7.2.3 that
$$M_r = e^r \int_r^\infty f(x) e^{-x} dx.$$

9. Prove by induction that if $f(X) = a_0 + a_1 X + \ldots + a_n X^n$ then
$$\int_0^r f(x) e^{-x} dx = \sum_{i=0}^n a_i i! - e^{-r} \sum_{i=0}^n f^{(i)}(r)$$
   for each $r \in \mathbb{R}$, where $f^{(i)}(r)$ is the i-th derivative of f evaluated at r. [Hint. For each integer m, take the induction statement to be
$$\int_0^r f(x) e^{-x} dx = \sum_{i=0}^m f^{(i)}(0) - e^{-r} \sum_{i=0}^m f^{(i)}(r) - \int_0^r f^{(m)}(x) e^{-x} dx.]$$

10. (a) Prove that e^r is irrational for all $r \in \mathbb{N}$.
    (b) Deduce from (a) and the formula just before (11) in the text that $\varepsilon_1, \ldots, \varepsilon_m$ are nonzero. Why didn't we need to use this fact in the proof of Theorem 7.2.8 (in contrast to the proof of Proposition 7.1.9 where ε_1 being nonzero is essential)?

7.3 Preliminaries on Symmetric Polynomials

The proof that π is transcendental (to be given in Section 7.6) is similar in many respects to that given in Section 7.2 for e. There are, however, two topics we must introduce before the proof, firstly symmetric polynomials (covered in this section) and secondly integrals of complex-valued functions (discussed in Section 7.5).

We are concerned here with polynomials in several indeterminates $X_1, X_2, \ldots, X_n$. Of course, $X_1 X_2^4 + 3X_3 = 3X_3 + X_2^4 X_1$, for example, since the order of multiplying indeterminates or adding terms makes no difference.

7.3.1 Example. With $n = 3$, two such polynomials are

$$f(X_1, X_2, X_3) = X_1^2 X_2^2 X_3^2 + 3X_1 + 3X_2 + 3X_3,$$
$$g(X_1, X_2, X_3) = X_1^2 X_2 + X_2^2 X_3.$$ ∎

Permutations and Symmetry

We are interested in what happens to such polynomials when the indeterminates are permuted amongst themselves, for example,

replacing X_1 by X_2, X_2 by X_3 and X_3 by X_1.

If we apply this permutation to f and g in the example above they become, respectively,

$$f_1(X_1, X_2, X_3) = X_2^2 X_3^2 X_1^2 + 3X_2 + 3X_3 + 3X_1,$$
$$g_1(X_1, X_2, X_3) = X_2^2 X_3 + X_3^2 X_1.$$

Notice that f and f_1 are equal polynomials but g and g_1 are not. You can see that, whatever permutation of the indeterminates is made in f, the resulting polynomial is equal to f. We say that f is a *symmetric polynomial* but that g is not. The precise definition is given as Definition 7.3.3 below.

Before giving that definition we need to say a little more about *permutations*. Intuitively a permutation of a set, for example the set $\{X_1, X_2, X_3\}$ of indeterminates above, is just a rearrangement of the order of the elements. More formally, a permutation ρ of a set S is a one-to-one and onto mapping $\rho : S \to S$. In the example above,

$$\rho(X_1) = X_2, \quad \rho(X_2) = X_3 \quad \text{and} \quad \rho(X_3) = X_1.$$

As you probably know, there are 6 ($= 3!$) permutations of any set such as $\{X_1, X_2, X_3\}$ with 3 elements. (We suggest you write them all down. Do not forget the so-called identity permutation which leaves all elements unchanged.) More generally there are $n!$ permutations of a set with n elements.

Clearly if f is a polynomial in the n indeterminates $X_1, X_2, \ldots, X_n$ and ρ is any permutation of the set, we get a new polynomial (which may be denoted by f^ρ) by applying ρ to all the X_i's in f.

7.3.2 Example. Let ρ_1 be the permutation of $\{X_1, X_2, X_3\}$ which interchanges X_1 and X_3 and leaves X_2 fixed, and let ρ_2 be the permutation given by $\rho_2(X_1) = X_3$, $\rho_2(X_2) = X_1$, $\rho_2(X_3) = X_2$. Then, with f and g as in Example 7.3.1 above,
$$g^{\rho_1}(X_1, X_2, X_3) = X_3^2 X_2 + X_2^2 X_1,$$
$$f^{\rho_2}(X_1, X_2, X_3) = X_3^2 X_1^2 X_2^2 + 3X_3 + 3X_1 + 3X_2.$$
You might like to write down $f^{\rho_1}(X_1, X_2, X_3)$ and $g^{\rho_2}(X_1, X_2, X_3)$. ∎

7.3.3 Definition. A polynomial $f(X_1, X_2, \ldots, X_n)$ in indeterminates $X_1, X_2, \ldots, X_n$ is called *symmetric* if, for all permutations ρ of $\{X_1, X_2, \ldots, X_n\}$, we have
$$f^\rho(X_1, X_2, \ldots, X_n) = f(X_1, X_2, \ldots, X_n).$$
∎

The next lemma contains two routine consequences of this definition. We leave the proof as Exercises 7.3 #7.

7.3.4 Lemma. (i) If $f(X_1, \ldots, X_n)$ and $g(X_1, \ldots, X_n)$ are symmetric polynomials in $X_1, \ldots, X_n$ then so are
$$f(X_1, \ldots, X_n) + g(X_1, \ldots, X_n),$$
$$f(X_1, \ldots, X_n) - g(X_1, \ldots, X_n),$$
$$f(X_1, \ldots, X_n)\, g(X_1, \ldots, X_n),$$
(their sum, difference, and product).

(ii) If $h(Y_1, \ldots, Y_m)$ is <u>any</u> polynomial in indeterminates $Y_1, \ldots, Y_m$ and if $g_1(X_1, \ldots, X_n), \ldots, g_m(X_1, \ldots, X_n)$ are symmetric polynomials in $X_1, \ldots, X_n$ then
$$h\bigl(g_1(X_1, \ldots, X_n), \ldots, g_m(X_1, \ldots, X_n)\bigr)$$
is also symmetric in $X_1, \ldots, X_n$. ∎

Note that "quotient" is not included in (i) of Lemma 7.3.4 since the quotient of two polynomials is in general not a polynomial.

Elementary Symmetric Functions

There is one way of obtaining several interesting and important symmetric polynomials which we illustrate in the case $n = 3$. Consider

$$(Y + X_1)(Y + X_2)(Y + X_3). \qquad (*)$$

If we expand this and collect the coefficients of the powers of the indeterminate Y we obtain

$$Y^3 + (X_1 + X_2 + X_3)Y^2 + (X_1X_2 + X_1X_3 + X_2X_3)Y + X_1X_2X_3.$$

Notice that the three polynomials obtained

$$\sigma_1(X_1, X_2, X_3) = X_1 + X_2 + X_3,$$
$$\sigma_2(X_1, X_2, X_3) = X_1X_2 + X_1X_3 + X_2X_3,$$
$$\sigma_3(X_1, X_2, X_3) = X_1X_2X_3,$$

are all symmetric. This is not surprising since clearly the value of the expression $(*)$ (from which they are derived) remains unchanged after any permutation of the X's. The three polynomials obtained are called the elementary symmetric functions in 3 indeterminates. This can be extended to any n in the following definition.

7.3.5 Definition. The n *elementary symmetric functions* $\sigma_1, \sigma_2, \ldots, \sigma_n$ *in indeterminates* $X_1, X_2, \ldots, X_n$ are the coefficients respectively of the powers $Y^{n-1}, Y^{n-2}, \ldots, Y^0$ in the expansion of

$$(Y + X_1)(Y + X_2) \ldots (Y + X_n);$$

that is,

$$\sigma_1(X_1, X_2, \ldots, X_n) = X_1 + X_2 + \ldots + X_n,$$
$$\sigma_2(X_1, X_2, \ldots, X_n) = X_1X_2 + X_1X_3 + \ldots + X_1X_n +$$
$$X_2X_3 + \ldots + X_2X_n +$$
$$\ldots + X_{n-1}X_n,$$
$$\vdots$$
$$\sigma_n(X_1, X_2, \ldots, X_n) = X_1X_2\ldots X_n. \qquad \blacksquare$$

Transcendence

7.3.6 Example.
(i) For $n=2$ we have $\sigma_1(X_1, X_2) = X_1 + X_2$ and $\sigma_2(X_1, X_2) = X_1 X_2$.
(ii) For $n=4$ we have

$$\sigma_1(X_1, X_2, X_3, X_4) = X_1 + X_2 + X_3 + X_4,$$
$$\sigma_2(X_1, X_2, X_3, X_4) = X_1 X_2 + X_1 X_3 + X_1 X_4 +$$
$$X_2 X_3 + X_2 X_4 + X_3 X_4,$$
$$\sigma_3(X_1, X_2, X_3, X_4) = X_1 X_2 X_3 + X_1 X_2 X_4 + X_1 X_3 X_4 + X_2 X_3 X_4,$$
$$\sigma_4(X_1, X_2, X_3, X_4) = X_1 X_2 X_3 X_4.$$

(iii) Note that, when $n = 4$, σ_1 has $\binom{4}{1}$ terms, σ_2 has $\binom{4}{2} = 6$ terms, σ_3 has $\binom{4}{3}$ terms and σ_4 has $\binom{4}{4}$ terms.
(iv) In general, $\sigma_k(X_1, X_2, \ldots, X_n)$ has $\binom{n}{k}$ terms. ∎

7.3.7 Remark. *The elementary symmetric functions provide a connection between the zeros of a polynomial and its coefficients.*

To see this connection, let

$$f(Y) = a_n Y^n + a_{n-1} Y^{n-1} + \ldots + a_1 Y + a_0 \qquad (**)$$

be a polynomial of degree n (that is, $a_n \neq 0$) with indeterminate Y and coefficients in a field $\mathbb{F}$. Assume also that f has n zeros $\gamma_1, \gamma_2, \ldots, \gamma_n$ in some (possibly larger) field $\mathbb{E}$ so that

$$f(Y) = a_n(Y - \gamma_1)(Y - \gamma_2) \ldots (Y - \gamma_n).$$

It follows immediately from the definition of the elementary symmetric functions (Definition 7.3.5) that this expands to

$$f(Y) = a_n Y^n - a_n \sigma_1(\gamma_1, \ldots, \gamma_n) Y^{n-1} + a_n \sigma_2(\gamma_1, \ldots, \gamma_n) Y^{n-2} -$$
$$\ldots + (-1)^n a_n \sigma_n(\gamma_1, \ldots, \gamma_n).$$

Thus from $(**)$

$$\sigma_1(\gamma_1, \ldots, \gamma_n) = -a_{n-1}/a_n,$$
$$\sigma_2(\gamma_1, \ldots, \gamma_n) = a_{n-2}/a_n,$$

and so on down to

$$\sigma_n(\gamma_1, \ldots, \gamma_n) = (-1)^n a_0/a_n.$$

Thus *each elementary symmetric function evaluated at the zeros of $(**)$ is plus or minus the quotient of a coefficient divided by the leading coefficient*. Note also that all of these quotients are in $\mathbb{F}$ (even though the γ's need not be in $\mathbb{F}$). ∎

The elementary symmetric functions are also important because, as we prove below in Theorem 7.3.10, every symmetric polynomial can be expressed in terms of (indeed, as a polynomial in) the elementary symmetric functions. The next example illustrates this.

7.3.8   Example. For f in Example 7.3.1,

$$f(X_1, X_2, X_3) = (X_1 X_2 X_3)^2 + 3(X_1 + X_2 + X_3) = h(\sigma_3, \sigma_1)$$

where $h(Y_1, Y_2) = Y_1^2 + 3Y_2$ and σ_3 and σ_1 are two of the elementary symmetric functions in the indeterminates X_1, X_2, X_3. ∎

In Example 7.3.8, it was convenient to write just σ_3 rather than $\sigma_3(X_1, X_2, X_3)$. Similarly, in what follows, we shall often omit the X's from the σ's. However, you should always remember that the σ's depend on indeterminates $X_1, X_2, \ldots, X_n$.

7.3.9   Definition. The *degree* of a monomial $X_1^{i_1} X_2^{i_2} \ldots X_n^{i_n}$, where $i_1, i_2, \ldots, i_n$ are non-negative integers, is defined as the sum of the exponents, $i_1 + i_2 + \ldots + i_n$, and the *degree* of any nonzero polynomial is the maximum degree of the monomials involved in it. ∎

In Example 7.3.8, note that the degree of h (which is 2) is no greater than the degree of f (which is 6), and that h has integer coefficients. The following theorem generalizes this example.

7.3.10   Theorem. [Fundamental Theorem on Symmetric Functions] *Every symmetric polynomial g, with coefficients in a field $\mathbb{F}$, in the indeterminates $X_1, X_2, \ldots, X_n$ can be written as a polynomial h (with coefficients in $\mathbb{F}$) in the n elementary symmetric functions. Moreover,*

(i) the degree of h is no more than the degree of g, and

(ii) if g has integer coefficients then so does h. ∎

Before proving the theorem, we describe an algorithm for constructing such a polynomial h. Later we shall prove that the algorithm works; this will prove the theorem. We first need another definition.

7.3.11 Definition. Consider two nonzero terms
$$cX_1^{i_1}X_2^{i_2}\ldots X_n^{i_n} \quad \text{and} \quad dX_1^{j_1}X_2^{j_2}\ldots X_n^{j_n} \quad (c,d \in \mathsf{F}).$$

We say that the first of these is of *higher order than* the second (or, equivalently that the second is of *lower order than* the first) if, at the first position, say k, in which the i's and j's differ, we have $i_k > j_k$. ∎

For example, $X_1^2 X_2^3 X_3$ is of higher order than each of $X_1^2 X_2 X_3$ and $X_1 X_2^5 X_3^{10}$, while $X_1 X_2^5 X_3^{10}$ is of lower order than $X_1^2 X_2^3 X_3$.

7.3.12 Algorithm. [Fundamental Algorithm for Symmetric Functions] The aim of the algorithm is as follows: Given a nonzero symmetric polynomial g in n indeterminates $X_1, X_2, \ldots, X_n$ satisfying the hypotheses of the Fundamental Theorem on Symmetric Functions, to construct a polynomial h satisfying the conclusion of that theorem.

<u>Step 0.</u> Initially set $\ell = 1$ and $g_1 = g$ (which is not zero).

<u>Step 1.</u>

(a) Find the term of highest order in g_ℓ, say
$$cX_1^{i_1}X_2^{i_2}\ldots X_n^{i_n} \quad (c \in \mathsf{F}, c \neq 0).$$

(b) With c and $i_1, i_2, \ldots, i_n$ as in (a), set
$$h_\ell(Y_1, Y_2, \ldots, Y_n) = cY_1^{i_1-i_2} Y_2^{i_2-i_3} \ldots Y_{n-1}^{i_{n-1}-i_n} Y_n^{i_n}.$$

(c) Put
$$g_{\ell+1}(X_1, X_2, \ldots, X_n) = g_\ell(X_1, X_2, \ldots, X_n) - h_\ell(\sigma_1, \sigma_2, \ldots, \sigma_n).$$

(d) Increase ℓ by 1.

(e) If $g_\ell \neq 0$ then go back to (a) and repeat Step 1; otherwise proceed to Step 2 below.

<u>Step 2.</u> Put $h = h_1 + h_2 + \ldots + h_{\ell-1}$. ∎

Before proving that this algorithm works, we give an example of its use.

7.3.13 Example. We apply the above algorithm to the symmetric polynomial (in $n = 2$ indeterminates)

$$g(X_1, X_2) = 3X_1^3 X_2 - 2X_1^2 X_2^2 + 3X_1 X_2^3.$$

(See Example 7.3.6 for formulae for σ_1 and σ_2.)

Step 0. Set $\ell = 1$ and put $g_1 = g$.

Step 1.
(a) The term of highest order in g_1 is $3X_1^3 X_2$.
(b) $h_1(Y_1, Y_2) = 3Y_1^{3-1} Y_2^1 = 3Y_1^2 Y_2$.
(c) $g_2(X_1, X_2) = g_1(X_1, X_2) - h_1(\sigma_1, \sigma_2)$
$= 3X_1^3 X_2 - 2X_1^2 X_2^2 + 3X_1 X_2^3 - 3(X_1 + X_2)^2 (X_1 X_2)$
$= -8X_1^2 X_2^2$.
(d) Set $\ell = 2$.
(e) As $g_2 \neq 0$, return to (a) and repeat Step 1 with g_2 in place of g_1.

Step 1.
(a) The term of highest order in g_2 is $-8X_1^2 X_2^2$.
(b) $h_2(Y_1, Y_2) = -8Y_1^{2-2} Y_2^2 = -8Y_2^2$.
(c) $g_3(X_1, X_2) = g_2(X_1, X_2) - h_2(\sigma_1, \sigma_2) = -8X_1^2 X_2^2 + 8X_1^2 X_2^2 = 0$.
(d) Set $\ell = 3$.
(e) Since $g_3 = 0$ go on to Step 2.

Step 2. Set $h = h_1 + h_2$; that is, $h(Y_1, Y_2) = 3Y_1^2 Y_2 - 8Y_2^2$.

As a check, note that

$$h(\sigma_1, \sigma_2) = 3(X_1 + X_2)^2 (X_1 X_2) - 8(X_1 X_2)^2$$
$$= 3X_1^3 X_2 - 2X_1^2 X_2^2 + 3X_1 X_2^3$$
$$= g(X_1, X_2),$$

and also that $\deg h = 3 < 4 = \deg g$ and that the polynomial h has integer coefficients. Thus h satisfies the conclusion of Theorem 7.3.10.■

To prove that the algorithm always works we need the following lemmas.

7.3.14 Lemma. Let $g(X_1, \ldots, X_n)$ be a symmetric polynomial over a field $\mathbb{F}$ and let

$$cX_1^{i_1} X_2^{i_2} \ldots X_n^{i_n} \quad (c \in \mathbb{F},\ c \neq 0) \tag{1}$$

be the term of highest order in g. Then

$$i_1 \geq i_2 \geq \ldots \geq i_n. \tag{2}$$

Proof. If, say, $i_k < i_{k+1}$ contrary to the conclusion of the lemma, then we apply to g the permutation ρ which just interchanges the indeterminates X_k and X_{k+1}. The resulting polynomial g^ρ (using the notation in Definition 7.3.3) then contains the term

$$cX_1^{i_1} \ldots X_{k-1}^{i_{k-1}} X_k^{i_{k+1}} X_{k+1}^{i_k} X_{k+2}^{i_{k+2}} \ldots X_n^{i_n} \tag{3}$$

which is of higher order than (1) by Definition 7.3.11. But g is a symmetric polynomial so the term (3) must be in $g = g^\rho$, which contradicts our choice of (1) as the term of highest order in g. Thus (2) holds. ∎

7.3.15 Lemma. The term of highest order in the expansion of the polynomial $\sigma_1{}^{i_1-i_2} \sigma_2{}^{i_2-i_3} \ldots \sigma_{n-1}{}^{i_{n-1}-i_n} \sigma_n{}^{i_n}$ is $X_1^{i_1} X_2^{i_2} \ldots X_n^{i_n}$.

Proof. The term of highest order in a product of polynomials is the product of the terms of highest order in each polynomial. The terms of highest order in $\sigma_1, \sigma_2, \ldots, \sigma_n$ are respectively

$$X_1,\ X_1 X_2,\ X_1 X_2 X_3,\ \ldots,\ X_1 X_2 X_3 \ldots X_n.$$

Hence the term of highest order in the product is

$$X_1{}^{i_1-i_2}(X_1 X_2)^{i_2-i_3} \ldots (X_1 X_2 \ldots X_n)^{i_n} = X_1^{i_1} X_2^{i_2} \ldots X_n^{i_n}. \quad \blacksquare$$

Proof of Theorem 7.3.10. To prove the theorem we prove that the Algorithm 7.3.12 always produces a polynomial h with the properties stated in the theorem.

If $g \neq 0$ is a symmetric polynomial in $X_1, X_2, \ldots, X_n$, we let $D \in \mathbb{N}$ denote the number of monomials (not necessarily occurring in g) which are of lower order than the highest term occurring in g. (Recall that a monomial is a term $X_1^{i_1} \ldots X_n^{i_n}$ with coefficient 1.) Our proof is by mathematical induction on D.

If $D = 0$ then g is a constant polynomial and it is left as an easy exercise to prove that the algorithm works in this case.

Now let $k \in \mathbb{N}$ and assume that the algorithm works for all symmetric polynomials g with $D \leq k$.

Consider now a polynomial g_1 with $D = k+1$, and let its term of highest order be

$$m_1 = cX_1^{i_1} X_2^{i_2} \ldots X_n^{i_n}, \tag{4}$$

where $c \in \mathbb{F}$ is nonzero.

⌈ We wish to prove that the algorithm works for g_1. ⌋

The first pass through Step 1 of the algorithm gives

$$h_1(Y_1, Y_2, \ldots, Y_n) = cY_1^{i_1-i_2} Y_2^{i_2-i_3} \ldots Y_n^{i_n} \tag{5}$$

and

$$g_2(X_1, X_2, \ldots, X_n) = g_1(X_1, X_2, \ldots, X_n) - h_1(\sigma_1, \sigma_2, \ldots, \sigma_n). \tag{6}$$

Now (5) defines h_1 as a polynomial since $i_1 \geq i_2 \geq \ldots \geq i_n$ by Lemma 7.3.14. Thus, by (ii) of Lemma 7.3.4, $h_1(\sigma_1, \ldots, \sigma_n)$ is symmetric in $X_1, \ldots, X_n$ and hence, by (i) of Lemma 7.3.4, so is g_2. By Lemma 7.3.15, moreover, the two polynomials on the right-hand side of (6) have the same term of highest order and so these terms cancel each other out in (6). If g_2 is zero, it is an easy exercise to check that the choice $h = h_1$ satisfies the requirements of the theorem. Hence we consider below only the case where $g_2 \neq 0$. If m_2 is the term of highest order in g_2, then m_2 is of lower order than m_1 (since the terms of highest order cancel each other out in (6)) and so the D for g_2 (the number of monomials of lower order than m_2) is less than $k+1$ which is the D for g_1 (the number of monomials of lower order than m_1). Thus the inductive assumption applies to g_2 and so the algorithm gives a polynomial h^*, say, such that

$$g_2(X_1, X_2, \ldots, X_n) = h^*(\sigma_1, \sigma_2, \ldots, \sigma_n).$$

Hence by (6), $g_1(X_1, X_2, \ldots, X_n) = h(\sigma_1, \sigma_2, \ldots, \sigma_n)$ where $h = h_1 + h^*$.

To see that the "moreover" parts of the theorem are true, note simply that

(i) the degree of $h_1(Y_1, \ldots, Y_n)$ in (5) is i_1 while the degree of g_1 is at least $i_1 + i_2 + \ldots + i_n$, so that the degree of h_1 is not more than the degree of g_1, and

Transcendence

(ii) if g_1 has integer coefficients then c in (4) is an integer, which means, by (5), that h_1 has integer coefficients.

The "moreover" results now follow easily. (Formally, each of these results requires a separate proof by induction.) ∎

7.3.16 Corollary. *Let F be a field and let $f(X)$ be a polynomial of degree n with coefficients in F and with n zeros $\alpha_1, \alpha_2, \ldots, \alpha_n$ in some extension field E of F. If g is any symmetric polynomial in $X_1, X_2, \ldots, X_n$ with coefficients in F then $g(\alpha_1, \alpha_2, \ldots, \alpha_n) \in$ F.*

Proof. By the Fundamental Theorem on Symmetric Functions, $g(X_1, X_2, \ldots, X_n)$ is equal to $h(\sigma_1, \sigma_2, \ldots, \sigma_n)$ where h is a polynomial with coefficients in F. It follows that

$$g(\alpha_1, \alpha_2, \ldots, \alpha_n) = h(\beta_1, \beta_2, \ldots, \beta_n)$$

where $\beta_i = \sigma_i(\alpha_1, \alpha_2, \ldots, \alpha_n)$ for $i = 1, 2, \ldots, n$. But, by Remark 7.3.7, each of the β's is in F (being the quotient of two of the coefficients of f) and thus $g(\alpha_1, \alpha_2, \ldots, \alpha_n)$ is in F. ∎

The next proposition plays a vital role in our proof (in Section 7.4 and Section 7.6) that π is transcendental.

7.3.17 Proposition. *Let F be a field and let $t(X)$ be a polynomial of degree n with coefficients in F and with n zeros $\alpha_1, \alpha_2, \ldots, \alpha_n$ in some extension field E of F. Assume that k is an integer between 1 and n, and let*

$$\gamma_1, \gamma_2, \ldots, \gamma_m$$

be all the sums of exactly k of the α's. Then there is a monic polynomial $t_k(X)$ of degree m with coefficients in F which has $\gamma_1, \gamma_2, \ldots, \gamma_m$ as its zeros.

7.3.18 Example. Let $n = 3$ and $k = 2$. Then there are three γ's, namely

$$\gamma_1 = \alpha_1 + \alpha_2, \quad \gamma_2 = \alpha_1 + \alpha_3, \quad \text{and} \quad \gamma_3 = \alpha_2 + \alpha_3.$$

It is clear that $t_2(X)$ must be

$$\begin{aligned} t_2(X) &= (X - \gamma_1)(X - \gamma_2)(X - \gamma_3) \\ &= X^3 - (\gamma_1 + \gamma_2 + \gamma_3)X^2 + (\gamma_1\gamma_2 + \gamma_1\gamma_3 + \gamma_2\gamma_3)X - \gamma_1\gamma_2\gamma_3. \end{aligned}$$

Each coefficient can be expressed in terms of the α's; for example, you can check that

$$\gamma_1\gamma_2 + \gamma_1\gamma_3 + \gamma_2\gamma_3 = \alpha_1^2 + \alpha_2^2 + \alpha_3^2 + 3(\alpha_1\alpha_2 + \alpha_1\alpha_3 + \alpha_2\alpha_3).$$

If you repeat this for other coefficients, you will see that each of the coefficients is a symmetric polynomial (with coefficients which are in $\mathbb{F}$ since they are integers) in the α's, and so is in $\mathbb{F}$ by Corollary 7.3.16. This is the essence of the proof of Proposition 7.3.17. ∎

Proof of Proposition 7.3.17. It is clear from the properties required of $t_k(X)$ that we must define it by

$$t_k(X) = (X - \gamma_1)(X - \gamma_2)\ldots(X - \gamma_m).$$

We must prove that each of its coefficients is in $\mathbb{F}$. Now, as in Remark 7.3.7, each of the coefficients of $t_k(X)$ is a symmetric polynomial evaluated at $(\gamma_1, \gamma_2, \ldots, \gamma_m)$. Thus it is sufficient to prove that, if h is any symmetric polynomial in m indeterminates and with coefficients in $\mathbb{F}$ then $h(\gamma_1, \gamma_2, \ldots, \gamma_m) \in \mathbb{F}$.

Accordingly, let h be any symmetric polynomial in m indeterminates and with coefficients in $\mathbb{F}$. We introduce n indeterminates $X_1, X_2, \ldots, X_n$ and let $Y_1, Y_2, \ldots, Y_m$ denote the m distinct sums of the X's taken k at a time. Then $h(Y_1, Y_2, \ldots, Y_m)$ can be expanded out to give a polynomial $g(X_1, X_2, \ldots, X_n)$ in the X's

$$h(Y_1, Y_2, \ldots, Y_m) = g(X_1, X_2, \ldots, X_n).$$

It is easy to see that if we permute the X's we also permute the Y's (since a sum of k X's remains such a sum after permutation of the X's, and every such sum comes from another such sum after permutation). Thus $g(X_1, X_2, \ldots, X_n)$ is a symmetric function in $X_1, X_2, \ldots, X_n$ and has its coefficients in $\mathbb{F}$. But, from the connection between the Y's and X's and between the γ's and α's, we have

$$h(\gamma_1, \gamma_2, \ldots, \gamma_m) = g(\alpha_1, \alpha_2, \ldots, \alpha_n)$$

which is in $\mathbb{F}$ by Corollary 7.3.16. ∎

Exercises 7.3

1. Write down formulae for the 6 permutations $\rho_1, \rho_2, \ldots, \rho_6$ of $\{X_1, X_2, X_3\}$. If
$$f(X_1, X_2, X_3) = X_1^2 X_2 + X_2^2 X_3,$$
write down $f^{\rho_i}(X_1, X_2, X_3)$ for $i = 1, 2, \ldots, 6$.

2. Write down two polynomials of degree 4 in X_1, X_2 which are symmetric and two which are not.

3. Assume that $f(X) = 2X^3 - 3X^2 + 4X - 1$ has three zeros $\alpha_1, \alpha_2, \alpha_3$ in $\mathbb{C}$. Calculate
$$\alpha_1 + \alpha_2 + \alpha_3, \quad \alpha_1\alpha_2 + \alpha_1\alpha_3 + \alpha_2\alpha_3, \quad \alpha_1\alpha_2\alpha_3.$$

4. Apply Algorithm 7.3.12 to express each of the following symmetric polynomials as polynomials in the relevant elementary symmetric functions.
   (a) $3X_1^3 X_2 + 6X_1^2 - 4X_1^3 X_2^3 + 12X_1 X_2 + 6X_2^2 + 3X_1 X_2^3$.
   (b) $5X_1^2 X_2^2 X_3 - 2X_1^2 X_2 X_3 - 2X_1 X_2^2 X_3 + 5X_1 X_2^2 X_3^2 + 5X_1^2 X_2 X_3^2 - 2X_1 X_2 X_3^2$.

5. Complete Example 7.3.18 by expressing the other coefficients of $t_2(X)$ in terms of the α's and checking that they are symmetric in the α's.

6. Consider the proof of Proposition 7.3.17 in the case $n = 4, k = 2$.
   (i) Write down the six γ's.
   (ii) Write down the coefficient of the X^5 term in
   $$t_2(X) = (X - \gamma_1)(X - \gamma_2) \ldots (X - \gamma_6).$$
   Express this in terms of the α's and check it is a symmetric function of the α's, as claimed in the proof of Proposition 7.3.17.

7. (a) Prove (i) of Lemma 7.3.4.
   [Hint. If ρ is any permutation of $\{X_1, \ldots, X_n\}$ then the result of applying ρ to $f(X_1, \ldots, X_n) + g(X_1, \ldots, X_n)$ is $f^\rho(X_1, \ldots, X_n) + g^\rho(X_1, \ldots, X_n)$.]

(b) Prove (ii) of Lemma 7.3.4. [Hint. If
$$t(X_1,\ldots,X_n) = h\Big(g_1(X_1,\ldots,X_n),\ldots,g_m(X_1,\ldots,X_n)\Big)$$
then, for any permutation ρ of $\{X_1,\ldots,X_n\}$,
$$t^\rho(X_1,\ldots,X_n) = h\Big(g_1^\rho(X_1,\ldots,X_n),\ldots,g_m^\rho(X_1,\ldots,X_n)\Big).]$$

(c) Can you see how to give an alternative proof of (ii) of Lemma 7.3.4, this time using (i) of Lemma 7.3.4?

7.4 π is Transcendental – Part 1

Complex Exponentials

In our proof that π is transcendental, we need to consider e^z where z is a complex number.

7.4.1 Definition. If $z = x + iy$ where $x, y \in \mathbb{R}$, then e^z is defined by
$$e^z = e^x(\cos(y) + i\sin(y)).$$
∎

7.4.2 Example.
$$e^{1+i} = e(\cos(1) + i\sin(1)),$$
$$e^{i\pi} = e^0(\cos\pi + i\sin\pi) = 1(-1 + 0i) = -1.$$
This latter fact will be the starting point for our proof that π is transcendental. It is also easy to check that $e^{z_1}e^{z_2} = e^{z_1+z_2}$ for all z_1 and z_2 in $\mathbb{C}$. (Use the expansions of $\cos(y_1+y_2)$ and $\sin(y_1+y_2)$.) ∎

An equation for π via symmetric polynomials

In the next proposition we need to use the well-known result that every polynomial with complex coefficients can be written as a product of factors of degree 1. More precisely, if $g(X) \in \mathbb{C}[X]$ has degree n and leading coefficient $c \in \mathbb{C}$ then $g(X)$ can be written as
$$g(X) = c(X - \lambda_1)(X - \lambda_2)\ldots(X - \lambda_n)$$
for complex numbers $\lambda_1, \lambda_2, \ldots, \lambda_n$ which are, of course, the zeros of $g(X)$ in $\mathbb{C}$. This result is known as the Fundamental Theorem of Algebra (see, for example, Theorem 6 in Chapter V, Section 3 of [GB]).

7.4.3 Proposition. *Suppose π is algebraic over $\mathbb{Q}$. Then there exist integers $m \geq 1$, $q \geq 1$, $b \geq 1$ and nonzero complex numbers $\beta_1, \beta_2, \ldots, \beta_m$ such that*

$$e^{\beta_1} + e^{\beta_2} + \ldots + e^{\beta_m} + q = 0 \qquad (1)$$

and the polynomial

$$h(X) = b(X - \beta_1)(X - \beta_2) \ldots (X - \beta_m) \qquad (2)$$

has integer coefficients.

Proof. Suppose π is algebraic.

> Because the equation $e^{i\pi} = -1$ (which relates π to an integer) has $i\pi$ rather than just π in the exponent, we consider the irreducible polynomial over $\mathbb{Q}$ of $i\pi$ rather than of π.

Since the complex number i is algebraic over $\mathbb{Q}$ it follows from Corollary 4.4.4 that the complex number $i\pi$ is algebraic (over $\mathbb{Q}$). If $t(X)$ denotes the irreducible polynomial of $i\pi$ over $\mathbb{Q}$ then $t(X)$ is a monic polynomial with rational coefficients such that $t(i\pi) = 0$. The Fundamental Theorem of Algebra tells us that $t(X)$ factors completely over $\mathbb{C}$ so that

$$t(X) = (X - \alpha_1)(X - \alpha_2) \ldots (X - \alpha_k)$$

for some complex numbers $\alpha_1, \alpha_2, \ldots, \alpha_k$. Of course one of these α's must be equal to $i\pi$ and we can suppose that $\alpha_1 = i\pi$.

But $e^{i\pi} = -1$ and hence

$$(e^{\alpha_1} + 1)(e^{\alpha_2} + 1) \ldots (e^{\alpha_k} + 1) = 0. \qquad (3)$$

> You may be puzzled as to why we include the terms $(e^{\alpha_2} + 1), \ldots, (e^{\alpha_k} + 1)$ in (3). These are essential because (as you will see later), by including the factors corresponding to <u>all</u> the zeros of f, we are able eventually to obtain symmetric polynomials and then integers.

If we multiply out the left-hand side (LHS) of (3), there are 2^k terms. If we use the fact that $e^{z_1} e^{z_2} = e^{z_1 + z_2}$ then we get $2^k - 1$ terms of the form e^γ (where γ is a sum of one or more α's) and a single term equal to 1 (the product of all the 1's in (3)).

> For example, if $n = 3$ the LHS of (3) is
> $$e^{\alpha_1+\alpha_2+\alpha_3} + e^{\alpha_1+\alpha_2} + e^{\alpha_1+\alpha_3} + e^{\alpha_2+\alpha_3} + e^{\alpha_1} + e^{\alpha_2} + e^{\alpha_3} + 1.$$

Thus (3) can be rewritten as
$$e^{\gamma_1} + e^{\gamma_2} + \ldots + e^{\gamma_N} + 1 = 0 \qquad (4)$$
(where $N = 2^k - 1$).

> Next we make our first use of results from Section 7.3 about symmetric polynomials.

Now we have met something like the formation of the γ's from the α's before in Proposition 7.3.17. Using that result and its notation,

(i) there is a monic polynomial $t_1(X)$ with rational coefficients which has all the α's as zeros,

(ii) there is a monic polynomial $t_2(X)$ with rational coefficients which has all the sums of two α's as zeros,

and so on, finishing with

(iii) a monic polynomial $t_k(X)$ with rational coefficients which has all the sums of k α's as zeros.

Thus if
$$T(X) = t_1(X)t_2(X)\ldots t_k(X)$$
then $T(X)$ is a monic polynomial with rational coefficients which has all the $N(= 2^k - 1)$ γ's as it zeros. This means that
$$T(X) = (X - \gamma_1)(X - \gamma_2)\ldots(X - \gamma_N). \qquad (5)$$

It may happen that some of these γ's (sums of α's) are in fact zero. (For example, it may happen that $\alpha_1 + \alpha_2 + \alpha_3 = 0$.) We have no way of knowing if this happens and, if so, how often it happens. But we can allow for it by saying that, if there are q_1 such γ's equal to zero, we can rewrite (4) as
$$e^{\beta_1} + e^{\beta_2} + \ldots + e^{\beta_m} + q = 0$$

Transcendence

where $\beta_1, \beta_2, \ldots, \beta_m$ are all nonzero complex numbers (the γ's which are nonzero) and $q = q_1 + 1$ is a positive integer, which proves (1). Note also that m must be at least 1 since, if $m = 0$, (1) would say that $q = 0$ which is impossible.

From (5)
$$T(X) = X^{q-1}(X - \beta_1)\ldots(X - \beta_m)$$
is a polynomial with rational coefficients and so
$$(X - \beta_1)\ldots(X - \beta_m) \tag{6}$$
is also a polynomial with rational coefficients. We can multiply the polynomial in (6) by a suitably large positive integer b (to cancel out all denominators there) to obtain a polynomial
$$h(X) = bX^m + b_{m-1}X^{m-1} + \ldots + b_0$$
$$= b(X - \beta_1)(X - \beta_2)\ldots(X - \beta_m)$$
with integer coefficients. (We have omitted the subscript from the leading coefficient b_m to simplify several formulae in Section 7.6.) ∎

Exercises 7.4

1. Prove from Definition 7.4.1 that $e^{z_1}e^{z_2} = e^{z_1+z_2}$ for all z_1 and $z_2 \in \mathbb{C}$.

2. If $z = x + iy$ where $x, y \in \mathbb{R}$, show that $|e^z| = e^x$.

7.5 Preliminaries on Complex-valued Integrals

Before we can complete the proof of the transcendence of π (in Section 7.6) we need to introduce integrals of the form
$$\int_a^b f(x)dx$$
where f is a *complex-valued function* (that is, a function $f : \mathbb{R} \to \mathbb{C}$) and a, b are real numbers. First we define integrals of this kind in terms of integrals of real-valued functions (such as you have learned about in your elementary calculus courses). We also define derivatives of complex-valued functions in terms of derivatives of real-valued functions.

7.5.1 Definition. Assume a, b are real numbers and $f : \mathbb{R} \to \mathbb{C}$ is a complex-valued function. For any $x \in \mathbb{R}$ we can write
$$f(x) = f_1(x) + i f_2(x)$$
where $f_1(x)$ and $f_2(x)$ are real numbers, which gives us two real-valued functions $f_1 : \mathbb{R} \to \mathbb{R}$ and $f_2 : \mathbb{R} \to \mathbb{R}$.

(i) We define
$$\int_a^b f(x)\,dx = \int_a^b f_1(x)\,dx + i \int_a^b f_2(x)\,dx$$
(provided the two normal real-valued integrals on the right-hand side exist).

(ii) We define the *derivative* $f'(x)$ as
$$f'(x) = f_1'(x) + i f_2'(x)$$
(provided the two real-valued derivatives on the right-hand side exist). ■

7.5.2 Example. Consider $f : \mathbb{R} \to \mathbb{C}$ given by
$$f(x) = (x + 2i)^2 \qquad \text{for } x \in \mathbb{R}.$$
Then
$$f(x) = (x^2 - 4) + 4xi$$
so that here $f_1(x) = x^2 - 4$ and $f_2(x) = 4x$. Thus, by the definitions above,
$$\int_1^2 (x + 2i)^2\,dx = \int_1^2 (x^2 - 4)\,dx + i \int_1^2 4x\,dx$$
$$= -\frac{5}{3} + 6i$$
and
$$f'(x) = f_1'(x) + i f_2'(x) = 2x + 4i. \qquad ■$$

Many of the properties of integrals and derivatives with which you are familiar carry over to these integrals and derivatives of complex-valued functions.* In the next proposition we collect some of the

---

* The subject of Complex Analysis, which is taught near the end of most undergraduate courses, covers integration and differentiation of more general functions $f : \mathbb{C} \to \mathbb{C}$. Because we expect that most of our readers will not yet have taken such a course, we have chosen to introduce in this section the special case $f : \mathbb{R} \to \mathbb{C}$, basing our development on material familiar from elementary (real) calculus courses. Of course, any readers familiar with complex analysis will find the material in this section easy special cases of what they already know.

properties we need below.

7.5.3   Proposition. *Let $f, g, F, h_1, \ldots, h_n : \mathbb{R} \to \mathbb{C}$ be complex-valued functions and let $a, b \in \mathbb{R}$ and $c, d_1, \ldots, d_n \in \mathbb{C}$. Then, provided the relevant integrals and derivatives below exist,*

(i) $\int_a^b cf(x)dx = c \int_a^b f(x)dx$;

(ii) $\int_a^b (f(x) + g(x))dx = \int_a^b f(x)dx + \int_a^b g(x)dx$;

(iii) [Linearity]
$$\int_a^b \Big(d_1 h_1(x) + \ldots + d_n h_n(x)\Big)dx = d_1 \int_a^b h_1(x)dx + \ldots + d_n \int_a^b h_n(x)dx$$;

(iv) [Fundamental Theorem of Calculus]

   if $F'(x) = f(x)$ then $\int_a^b f(x) = F(b) - F(a)$;

(v) [Integration by Parts]
$$\int_a^b f(x)g'(x)dx = f(b)g(b) - f(a)g(a) - \int_a^b f'(x)g(x)dx \, .$$

Proof.   First we do (i) in some detail. Let $f(x) = f_1(x) + if_2(x)$ and $c = c_1 + c_2 i$, where $f_1(x), f_2(x), c_1$ and c_2 are real. Then

$$cf(x) = (c_1 f_1(x) - c_2 f_2(x)) + i(c_1 f_2(x) + c_2 f_1(x))$$

and so, by Definition 7.5.1,

$$\int_a^b cf(x)dx = \int_a^b (c_1 f_1(x) - c_2 f_2(x))dx + i \int_a^b (c_1 f_2(x) + c_2 f_1(x))dx.$$

The right-hand side of (i) is easily seen to be the same. Part (ii) follows similarly, while (iii) follows easily from (i) and (ii).

The remaining results also follow easily from Definition 7.5.1 and the corresponding properties of real-valued functions. We suggest you write out the details of at least some of these other parts (see Exercises 7.5 #3 below). ∎

7.5.4   Lemma. *Let $r \in \mathbb{C}$ and consider $g : \mathbb{R} \to \mathbb{C}$ given by $g(u) = e^{-ur}$ for $u \in \mathbb{R}$. Then $g'(u) = -re^{-ur}$ for all $u \in \mathbb{R}$.*

Proof.   If $r = r_1 + r_2 i$ (r_1, r_2 real) then, from Definition 7.4.1,

$$g(u) = e^{-ur_1}(\cos(-ur_2) + i\sin(-ur_2))$$

and the result follows easily from the definition of derivative in Definition 7.5.1. ∎

We also need the following lemma which gives an upper bound for the modulus of an integral. (Recall that the *modulus* $|z|$ of the complex number $z = c + di$ is given by $|z| = \sqrt{c^2 + d^2}$.)

7.5.5 Lemma. *Assume* $f : \mathbb{R} \to \mathbb{C}$ *and* $a, b, c \in \mathbb{R}$ *are such that* $|f(x)| \leq c$ *for all* $x \in [a, b]$ *where* $a < b$. *Then*

$$\left| \int_a^b f(x) dx \right| \leq 2c(b - a).$$

Proof. Let $f(x) = f_1(x) + i f_2(x)$, where $f_1(x)$ and $f_2(x)$ are real. Then, if $x \in [a, b]$,

$$|f_1(x)| \leq \sqrt{f_1(x)^2 + f_2(x)^2} = |f(x)| \leq c$$

and similarly $|f_2(x)| \leq c$. Now, from (i) of Definition 7.5.1,

$$\left| \int_a^b f(x) dx \right| = \sqrt{\left(\int_a^b f_1(x) dx \right)^2 + \left(\int_a^b f_2(x) dx \right)^2}$$
$$\leq \left| \int_a^b f_1(x) dx \right| + \left| \int_a^b f_2(x) dx \right| \quad (*)$$
$$\leq c(b - a) + c(b - a) \quad (**)$$
$$= 2c(b - a)$$

as required. (Note that $(*)$ follows because, as is easily seen by squaring, $\sqrt{r^2 + s^2} \leq |r| + |s|$ for all $r, s \in \mathbb{R}$, while $(**)$ follows because $|f_1(x)| \leq c$ and $|f_2(x)| \leq c$.) ∎

──────────────── **Exercises 7.5** ────────────────

1. Evaluate the complex-valued integral $\int_0^1 (x + i)^3 dx$.

2. If $f(x) = (x + i)^3$ for all $x \in \mathbb{R}$, use (ii) of Definition 7.5.1 to calculate $f'(x)$.

3. Complete the proof of Proposition 7.5.3, being careful to indicate which properties of integrals of real-valued functions you use.

4. Complete the proof of Lemma 7.5.4.

5. Let $f : \mathbb{R} \to \mathbb{C}$ be given by $f(u) = ue^{-ui}$ for all $u \in \mathbb{R}$. Use integration by parts to calculate $\int_0^1 f(u) du$.

7.6 π is Transcendental – Part 2

Proposition 7.4.3 is the beginning of our proof that π is transcendental. Recall that we proved there that, if π is algebraic, then there exist integers $m \geq 1$, $q \geq 1$, $b \geq 1$ and nonzero complex numbers $\beta_1, \beta_2, \ldots, \beta_m$ such that

$$e^{\beta_1} + e^{\beta_2} + \ldots + e^{\beta_m} + q = 0 \tag{1}$$

and the polynomial

$$h(X) = b(X - \beta_1)(X - \beta_2) \ldots (X - \beta_m) \tag{2}$$

has integer coefficients.

Multiplying out $h(X)$ gives

$$h(X) = b_0 + b_1 X + \ldots + b_{m-1} X^{m-1} + b X^m \tag{3}$$

where $b_0, \ldots, b_{m-1}, b$ are integers and the constant term

$$b_0 = (-1)^m b \beta_1 \beta_2 \ldots \beta_m$$

is nonzero since the b and β's are all nonzero.

Now that the above equations have been obtained, the rest of our proof that π is transcendental is going to look very much like the proof for e given in Section 7.2. We are going to derive a contradiction from (1) just as we derived a contradiction from the corresponding result ((1) of Section 7.2) for e. To this end we shall show how to construct certain numbers $M, M_{\beta_1}, \ldots, M_{\beta_m}$ and $\varepsilon_{\beta_1}, \ldots, \varepsilon_{\beta_m}$ by using integrals, as we did in Section 7.2 in the proof for e.

The integrals we shall use have the form

$$\int_0^1 f(ur) e^{-ur} r \, du \tag{$*$}$$

where r is a <u>complex number</u>. Interested readers might like to compare this with the integrals

$$\int_0^r f(x) e^{-x} dx \qquad (r \in \mathbb{R}) \tag{$**$}$$

we used in Section 7.2. Although it is not used in our proof below, you may be interested to note that if we make the formal substitution $x = ur$ in $(**)$ we obtain $(*)$.

Producing the *M*'s and the ε's

We state and prove below four lemmas needed to define the *M*'s and the ε's and to derive their properties. Each of these lemmas is very similar to one of the lemmas in Section 7.2.

The first lemma is the analogue of Lemma 7.2.1.

7.6.1 Lemma. *For each integer $k \geq 0$ there are polynomials $g_k(X), h_k(X)$ with real coefficients such that, for all $r \in \mathbb{C}$,*

(a) $$\int_0^1 (ur)^k e^{-ur} r\, du = k! - e^{-r} g_k(r),$$

(b) $$\int_0^1 (ur - r)^k e^{-ur} r\, du = h_k(r) - e^{-r} k!.$$

Proof. This can be proved as for Lemma 7.2.1 by mathematical induction on k, using the relevant integration by parts rule (see (v) of Proposition 7.5.3). Lemma 7.5.4 gives the rule needed for differentiating (or antidifferentiating) e^{-ur} with respect to u. ∎

The next lemma is the analogue of Lemma 7.2.3.

7.6.2 Lemma. *Given a polynomial $f(X)$ with real coefficients, there is a unique number M and a unique polynomial $G(X)$ with real coefficients such that, for all $r \in \mathbb{C}$,*

$$\int_0^1 f(ur) e^{-ur} r\, du = M - e^{-r} G(r).$$

Indeed M and $G(X)$ are unique in the strong sense that, if $P_1(X)$ and $P_2(X)$ are polynomials with real coefficients such that

$$\int_0^1 f(ur) e^{-ur} r\, du = P_1(r) - e^{-r} P_2(r)$$

for all $r \in \mathbb{C}$, then $P_2(X) = G(X)$ and $P_1(X) = M$.

Proof. The proof is very similar to that of Lemma 7.2.3. For the existence part, use linearity as in (iii) of Proposition 7.5.3 instead of (i) of Theorem 7.1.2, and then use Lemma 7.6.1(a) instead of Lemma 7.2.1(a). The uniqueness part follows the proof in Lemma 7.2.3 exactly since in this case

$$M - P_1(r) = e^{-r}\Big(G(r) - P_2(r)\Big)$$

holds for all $r \in \mathbb{C}$ and so, in particular, for all $r \in \mathbb{R}$. ∎

Transcendence

As in the proof for e, our choice of the numbers M etc involves using a particular choice of polynomial in the integrand.

7.6.3 Definition. In Lemma 7.6.2 we choose $f(X)$ to be the polynomial
$$f(X) = \frac{X^{p-1}\bigl(h(X)\bigr)^p}{(p-1)!} \qquad (4)$$
where the polynomial $h(X)$ is given by (2) and (3) and where p is a prime bigger than the integers q, b and $|b_0|$ occurring in (1), (2) and (3). We let $n = mp + p - 1$, which is the degree of $f(X)$.

We then choose

(i) M and $G(X)$ to be the number and polynomial given in Lemma 7.6.2,

(ii) $M_r = b^n G(r)$ for $r \in \{\beta_1, \ldots, \beta_m\}$, and

(iii) $\varepsilon_r = b^n e^r \int_0^1 f(ur) e^{-ur} r\, du$ for $r \in \{\beta_1, \ldots, \beta_m\}$. ∎

Properties of $M, M_{\beta_1}, \ldots, M_{\beta_m}$ and $\varepsilon_{\beta_1}, \ldots, \varepsilon_{\beta_m}$

The next lemma is the analogue of Lemma 7.2.6.

7.6.4 Lemma. *(a) The number M is an integer which does not have the prime p as a factor when $p > |b_0|$, where b_0 is as in (3) above.*

(b) There is a polynomial $G_1(X)$ with integer coefficients and of degree at most n such that $G(r) = p\,G_1(r)$ for $r \in \{\beta_1, \ldots, \beta_m\}$.

Proof. The proof of each part (a) and (b) is very similar to the corresponding part of Lemma 7.2.6. Here we merely indicate the changes that are required and leave the details as Exercises 7.6 #3.

To prove (a), replace x by ur and dx by du, and use Lemmas 7.6.1(a) and 7.6.2 in place of Lemmas 7.2.1(a) and 7.2.3. Recall that $b_0 \ne 0$ in (3).

To prove (b), firstly expand $f(X)$ in powers of $(X - r)$, with $r \in \mathbb{C}$, using the obvious complex analogue of Proposition 7.1.6. Again replace x by ur and use Lemmas 7.6.1(b) and 7.6.2 in place of Lemmas 7.2.1(b) and 7.2.3. That each of $d_0(r), \ldots, d_{p-1}(r)$ is zero if r is in $\{\beta_1, \beta_2, \ldots, \beta_m\}$ follows just as in Lemma 7.2.5, using the complex analogues of Proposition 7.1.6 and Lemma 7.1.7. ∎

> Note a significant difference between Lemma 7.2.6(b) and Lemma 7.6.4(b). In the latter we make no claim that $G_1(r)$ is an integer for $r \in \{\beta_1, \ldots, \beta_m\}$. Although $G_1(X)$ has integer coefficients, the β's are complex numbers and so we do not expect $G_1(r)$ to be an integer (unlike Section 7.2 where $r \in \{1, 2, \ldots, m\}$ so r is an integer, guaranteeing that $G_1(r) \in \mathbb{Z}$). In the proof for π we shall use symmetric polynomials to show that the sum
>
> $$G_1(\beta_1) + \ldots + G_1(\beta_m)$$
>
> is an integer (even though the individual $G_1(\beta_i)$'s may not be).

Note that a sequence $\{z_n\}$ of complex numbers converges to 0 as $n \to \infty$ if the sequences of real and imaginary parts of z_n converge to 0; this is equivalent to saying that the sequence $\{|z_n|\}$ of moduli converges to 0.

Our final lemma is the analogue of Lemma 7.2.7.

7.6.5 Lemma. *Let $f(X)$ and n be as in Definition 7.6.3. For $r \in \{\beta_1, \ldots, \beta_m\}$ let ε_r be as in Definition 7.6.3, so that*

$$\varepsilon_r = b^n e^r \int_0^1 f(ur) e^{-ur} r \, du.$$

Then

$$\lim_{p \to \infty} \varepsilon_r = 0.$$

Proof. Let $u \in [0, 1]$ and let $x = ur$ where r is the complex number $r = r_1 + ir_2$ (where $r_1, r_2 \in \mathbb{R}$). By definition,

$$e^r = e^{r_1}(\cos r_2 + i \sin r_2)$$

and so $|e^r| = e^{r_1}$. Similarly $|e^{-ur}| = e^{-ur_1} \leq e^{|r_1|}$. Also, from (4) and (2),

$$|f(x)| = \left| \frac{x^{p-1}}{(p-1)!} b^p (x - \beta_1)^p \ldots (x - \beta_m)^p \right|$$

$$\leq \frac{|r|^{p-1}}{(p-1)!} b^p \bigl(|r| + |\beta_1|\bigr)^p \ldots \bigl(|r| + |\beta_m|\bigr)^p.$$

Hence, using Lemma 7.5.5,

$$\left| b^n e^r \int_0^1 f(ur)e^{-ur} r\, du \right|$$

$$\leq b^{mp+p-1} e^{r_1} (2 \text{ length of interval}) \times (\text{upper bound for } |f(ur)e^{-ur}r|)$$

$$\leq b^{mp+p-1} e^{r_1} \cdot 2 \cdot 1 \cdot \frac{|r|^{p-1} b^p}{(p-1)!} \Big(|r|+|\beta_1|\Big)^p \cdots \Big(|r|+|\beta_m|\Big)^p e^{|r_1|} |r|$$

$$\leq \frac{2 e^{r_1+|r_1|}}{b} \cdot \frac{C^p}{(p-1)!}$$

where $C = b^{m+2}|r|\Big(|r|+|\beta_1|\Big) \cdots \Big(|r|+|\beta_m|\Big)$.

Note that C is independent of p (and so are b and r_1). Hence by Lemma 7.1.4, $\lim_{p \to \infty} \varepsilon_r = 0$. ■

Given the above results, the long awaited proof of the transcendence of π is relatively straightforward.

7.6.6 Theorem. *π is a transcendental number.*

Proof. Suppose π is algebraic. We shall derive a contradiction.

We have already proved various consequences of this assumption and restated them at the start of this section. We let p be any prime which is larger than all of q, b and $|b_0|$ occurring in (1), (2) and (3), and define the polynomial $f(X)$ and the integer n as in (4) above.

It follows from Lemma 7.6.4 that there is an integer M not divisible by p and a polynomial $G_1(X)$ with integer coefficients and degree at most n such that, for $r \in \{\beta_1, \beta_2, \ldots, \beta_m\}$,

$$\int_0^1 f(ur)e^{-ur}r\, du = M - e^{-r} p G_1(r). \tag{5}$$

We now introduce the numbers $M_{\beta_1}, \ldots, M_{\beta_m}$ and $\varepsilon_{\beta_1}, \ldots, \varepsilon_{\beta_m}$.

Let M and M_r, ε_r, for each r in $\{\beta_1, \beta_2, \ldots, \beta_m\}$, be as in Definition 7.6.3. From (iii) of Definition 7.6.3 and Lemma 7.6.4(b),

$$M_r = p b^n G_1(r) \tag{6}$$

and by (iii) of Definition 7.6.3,
$$\varepsilon_r = b^n e^r \int_0^1 f(ur) e^{-ur} r \, du. \tag{7}$$

Then, from (5), (6) and (7), we have
$$M_r + \varepsilon_r = p b^n G_1(r) + b^n e^r (M - e^{-r} p G_1(r)) = b^n e^r M$$

and so
$$e^r = \frac{M_r + \varepsilon_r}{M b^n} \quad \text{for } r \in \{\beta_1, \beta_2, \ldots, \beta_m\}. \tag{8}$$

> Unlike the case for e, here the numbers $M_{\beta_1}, \ldots, M_{\beta_m}$ may not be integers – indeed all that we can be sure of is that they are complex numbers since the β's and hence the r in (6) are complex. We only get the integer we need by adding them all up and then recognizing that there is a symmetric polynomial involved.

Now $G_1(X_1) + G_1(X_2) + \ldots + G_1(X_m)$ is clearly a symmetric polynomial in $X_1, X_2, \ldots, X_m$ and so it follows from the Fundamental Theorem on Symmetric Functions (Theorem 7.3.10) that
$$G(X_1) + G(X_2) + \ldots + G(X_m) = H(\sigma_1, \sigma_2, \ldots, \sigma_m)$$
where H is a polynomial with integer coefficients and degree at most n. Substituting β_i for X_i ($i = 1, 2, \ldots, m$) we see that
$$G_1(\beta_1) + G_1(\beta_2) + \ldots + G_1(\beta_m) = H(\mu_1, \mu_2, \ldots, \mu_m) \tag{9}$$
where μ_i is the i-th elementary symmetric function σ_i evaluated at $(\beta_1, \beta_2, \ldots, \beta_m)$. But, by (2) and (3) and Remark 7.3.7,
$$\mu_i = \pm b_{m-i}/b.$$

Since the degree of H is at most n, each term on the right-hand side of (9) is equal to an integer divided by some power (no bigger than n) of b (since $b_0, b_1, \ldots, b_{m-1}$ and b are all integers). Thus
$$G_1(\beta_1) + G_1(\beta_2) + \ldots + G_1(\beta_m) = d/b^n$$
for some integer d, and so, from (6),
$$M_{\beta_1} + M_{\beta_2} + \ldots + M_{\beta_m} = p b^n (d/b^n) = pd$$
is an integer divisible by p.

Transcendence

> Now that we know that this sum is an integer with p as a factor, we can choose p large enough to get a contradiction in essentially the same way as in the proof for e.

Substituting (8) into (1) gives

$$[qMb^n + M_{\beta_1} + \ldots + M_{\beta_m}] + [\varepsilon_{\beta_1} + \ldots + \varepsilon_{\beta_m}] = 0. \qquad (10)$$

But, since the integer M is not divisible by the prime p and since p is larger than the integers q and $|b|$, it follows that qMb^n is an integer not divisible by p. This means that the term in the first [ ] of (10) is an integer not divisible by p and hence is a nonzero integer. On the other hand, by Lemma 7.6.5, the modulus of each term inside the second [ ] of (10) can be made arbitrarily small by choosing p large enough, and so the modulus of this second [ ] term in (10) can be made less than $\frac{1}{2}$ (for example) by choosing p large enough. This is the contradiction we promised.

Hence π is transcendental. ∎

7.6.7 Corollary. *Squaring the circle is not possible with straightedge and compass.*

Proof. This follows immediately from Theorem 7.6.6 and the remarks in Section 6.2.3. ∎

――――――― **Exercises 7.6** ―――――――

1. Complete the details of the proof of Lemma 7.6.1, being careful to check which properties from Section 7.5 you use.

2. Repeat Exercise 1 for Lemma 7.6.2.

3. Complete the details of the proof of Lemma 7.6.4. State and prove the complex analogues of Propositions 7.1.6 and Lemma 7.1.7 which you need. Prove the analogue of Lemma 7.2.5 which you need.

Additional Reading for Chapter 7

The idea of using integration to prove that e is transcendental was introduced in Hermite's original proof in [ChH]. This proof for e was extended by Lindemann in [FL] to give a proof that π is transcendental. Simplified versions of the original proofs have been given by many authors; except for [FB], they all use the polynomials $f(X)$ given in Definitions 7.2.4 and 7.6.3. They all use results about symmetric polynomials, moreover, in the proofs for π.

Two variants of the proof stem from Hilbert in [DH] on the one hand, and from Hurwitz in [AH] on the other. Although Hurwitz gives only the proof for e, his method was extended later to π by Niven in [IN3]. In Hilbert's approach, the integral of $f(x)e^{-x}$ is obtained by first expanding the polynomial $f(x)$ in suitable powers, and then integrating term by term. This seems essentially easier than Hurwitz's approach in which the integral is expressed as the sum of the successive derivatives of $f(x)$, left in factorized form. In the proof for π, however, Niven's approach is more elementary than Hilbert's in that it avoids use of integration in the complex plane (and the Cauchy integral theorem). Our proof is an attempt to combine the advantages of both of these approaches.

References which follow Hilbert's approach are [AB], [FK1] and [MS] – although the last gives only the proof for e. References which follow the Hurwitz-Niven approach include [CH], [IN2] and [IN3].

In an attempt to make the proof even more elementary, a number of authors have removed integrals entirely from their proofs and replaced them by sums, as in [PG], [EH], [GH], [FK] and [DS]. Spivak remarks that such elementary proofs are often harder to understand than the original proofs, inasmuch as the essential structure of the proof has been obscured just to remove a few integral signs. What do readers think?

An approach to symmetric functions which is similar to ours is contained in [WF] and [CH]. A proof of the Fundamental Theorem on Symmetric Functions which uses a double induction, rather than the concept of order of a monomial, is given in [JA] and [AC].

[JA] J. Archbold, *Algebra*, 4th edition, Pitman, London, 1970.

[AB] A. Baker, *Transcendental Number Theory*, Cambridge University Press, Cambridge, 1975.

[FB] F. Beukers, J.P. Bezivia and P. Robba, "An Alternative Proof of the Lindemann – Weierstrass Theorem", *American Mathematical Monthly*, 97 (1990), 193-197.

[GB] G. Birkhoff and S. MacLane, *A Survey of Modern Algebra*, Macmillan, New York, 1953.

[AC] A. Clark, *Elements of Abstract Algebra*, Wadsworth, Belmont, California, 1971.

[WF] W.L. Ferrar, *Higher Algebra*, Clarendon, Oxford, 1958.

[PG] P. Gordan, "Transcendenz von e und π", *Mathematische Annalen*, 43 (1893), 222-224.

[AH] A. Hurwitz, "Beweis der Transcendenz der Zahl e", *Mathematische Annalen*, 43 (1893), 220-222.

[CH] C.R. Hadlock, *Field Theory and its Classical Problems*, Carus Mathematical Monographs, No. 19, Mathematical Association of America, 1978.

[ChH] Ch. Hermite, "Sur la fonction exponentielle", *Comptes Rendus des Séances de l'Académie des Sciences Paris*, 77 (1873), 18- 24.

- [DH] D. Hilbert, "Über die Transcendenz der Zahlen e und π", *Mathematische Annalen*, 43 (1893), 216-219; reprinted in *Gesammelte Abhandlungen Vol.1*, Chelsea, 1965.
- [EH] E.W. Hobson, *Squaring the Circle*, Cambridge University Press, 1913; reprinted in *Squaring the Circle and Other Monographs*, Chelsea, 1953.
- [GH1] G.H. Hardy, *A Course of Pure Mathematics*, 10th edition, Cambridge University Press, Cambridge, 1952.
- [GH] G.H. Hardy and E.M. Wright, *An Introduction to the Theory of Numbers*, 3rd edition, Clarendon, Oxford, 1954.
- [FK1] F. Klein, *Elementary Mathematics from an Advanced Standpoint*, (vol 1: Arithmetic, Algebra and Analysis), Dover, New York, 1948.
- [FK] F. Klein, *Famous Problems of Elementary Geometry*; reprinted in *Famous Problems and Other Monographs*, Chelsea, 1962.
- [FL] F.L. Lindemann, "Über die Zahl π", *Mathematische Annalen*, 20 (1882), 213-225.
- [IN2] I. Niven, *Irrational Numbers*, Carus Mathematical Monographs, No.11, Mathematical Association of America, 1963.
- [IN3] I. Niven, "The transcendence of π", *American Mathematical Monthly*, 46 (1939), 469-471.
- [DS] D.E. Smith, *The History and Transcendence of π*; reprinted in W.A. Young, *Monographs on Topics of Modern Mathematics Relevant to the Elementary Field*, Dover, 1955.
- [MS] M. Spivak, *Calculus*, Benjamin, New York, 1967.

CHAPTER 8

An Algebraic Postscript

We have completed the proof that the three famous constructions are impossible. In this chapter we introduce some purely algebraic results which extend results in the earlier chapters. These new results will enable you to construct fields that you have not seen before.

If α is a complex number which is algebraic over a subfield $\mathbb{F}$ of $\mathbb{C}$, we have been able to show that $\mathbb{F}(\alpha)$ is a field, but we have, as yet, introduced no effective way of calculating reciprocals of numbers in $\mathbb{F}(\alpha)$. In this chapter, we describe a way of constructing fields which gives, as a by-product, an algorithm for calculating such reciprocals.

The material in this chapter is optional as it is, to some extent, a digression from our main theme of "impossibilities". We return to this theme in the short concluding Chapter 9.

8.1 The Ring $\mathbb{F}[X]_{p(X)}$

We describe a set $\mathbb{F}[X]_{p(X)}$ of polynomials and define addition and multiplication operations for these polynomials. It will turn out that $\mathbb{F}[X]_{p(X)}$ is a field with these operations.

8.1.1 Definitions. Let $\mathbb{F}$ be any field and let $p(X)$ be a polynomial in $\mathbb{F}[X]$ which is irreducible over $\mathbb{F}$ and which has degree n. Define the set $\mathbb{F}[X]_{p(X)}$ by putting

$$\mathbb{F}[X]_{p(X)} = \{a_0 + a_1 X + \ldots + a_{n-1} X^{n-1} : a_0, a_1, \ldots, a_{n-1} \in \mathbb{F}\}.$$

We also define addition and multiplication in $\mathbb{F}[X]_{p(X)}$: for each $f(X)$ and $g(X)$ in $\mathbb{F}[X]_{p(X)}$, define

(i) $f(X) \oplus_{p(X)} g(X) = f(X) + g(X)$,

(ii) $f(X) \otimes_{p(X)} g(X) =$ remainder on dividing $f(X)g(X)$ by $p(X)$. ■

Notice that, as a set, $\mathbb{F}[X]_{p(X)}$ depends only on the degree n of the polynomial $p(X)$. The set $\mathbb{F}[X]_{p(X)}$ contains all polynomials in $\mathbb{F}[X]$ of degree less than n (together with the zero polynomial). Addition $\oplus_{p(X)}$ in $\mathbb{F}[X]_{p(X)}$ is just ordinary addition of polynomials while multiplication $\otimes_{p(X)}$ depends on the polynomial $p(X)$. Indeed, multiplication is done *modulo* $p(X)$ in the same way that multiplication $\otimes_n$ in $\mathbb{Z}_n$ is done modulo n (as described in Section 1.1). The usual product $f(X) g(X)$ may have too high a degree for it to belong to $\mathbb{F}[X]_{p(X)}$; to bring down the degree we remove multiples of $p(X)$.

8.1.2 Example. In the set of polynomials $\mathbb{Q}[X]_{X^3-2}$ we have

$$(X^2+3) \oplus_{X^3-2} (2X+1) = X^2 + 2X + 4$$

and

$$(X^2+3) \otimes_{X^3-2} (2X+1) = X^2 + 6X + 7$$

since $(X^2+3)(2X+1) = 2X^3 + X^2 + 6X + 3$ which leaves a remainder of $X^2 + 6X + 7$ when divided by $X^3 - 2$. ■

8.1.3 Proposition. $(\mathbb{F}[X]_{p(X)}, \oplus_{p(X)}, \otimes_{p(X)})$ *is a ring.*

Proof. Addition is just ordinary addition while multiplication is closely related to ordinary multiplication. The proof of the various ring axioms follows closely the arguments which one would use to prove that $(\mathbb{Z}_n, \oplus_n, \otimes_n)$ is a ring. The details are left for you to consider. ■

An Algebraic Postscript

In fact $F[X]_{p(X)}$ is always a ring, even if $p(X)$ is reducible over F. The irreducibility of $p(X)$ over F is only needed (as we shall see in the next section) to prove that $F[X]_{p(X)}$ is a field. This is also analogous to $\mathbb{Z}_n$, which is always a ring, but only a field if n is a prime number.

Notice that all the constant polynomials are in $F[X]_{p(X)}$ so that $F \subseteq F[X]_{p(X)}$. Indeed $F[X]_{p(X)}$ is obviously a vector space over F with a basis $\{1, X, \ldots, X^{n-1}\}$ and dimension n.

──────────── **Exercises 8.1** ────────────

1. In $\mathbb{Q}[X]_{X^3-2}$, calculate
    (a) $(X^2 - 3X + 2) \oplus_{X^3-2} (2X^2 + X + 4)$,
    (b) $(X^2 - 3X + 2) \otimes_{X^3-2} (2X^2 + X + 4)$.

2. (a) Explain why $X^2 + 1$ is irreducible over $\mathbb{R}$.
    (b) In $\mathbb{R}[X]_{X^2+1}$, calculate
        (i) $(3+4X) \oplus_{X^2+1} (2-X)$,   (ii) $(3+4X) \otimes_{X^2+1} (2-X)$,
        (iii) $(a+bX) \oplus_{X^2+1} (c+dX)$,   (iv) $(a+bX) \otimes_{X^2+1} (c+dX)$.
    (c) Do the formulae in (b) (iii), (iv) look familiar? (If not, write down formulae for addition and multiplication of two complex numbers.)

3. (a) Show that in the ring $\mathbb{Q}[X]_{X^3-1}$, the element $X - 1$ is a zero divisor. [Hint. Write $X^3 - 1$ as a product of two polynomials of smaller degree, and then multiply those polynomials modulo $X^3 - 1$.]
    (b) Is $X^3 - 1$ irreducible over $\mathbb{Q}$? Is $\mathbb{Q}[X]_{X^3-1}$ a field?

8.2 Division and Reciprocals in $F[X]_{p(X)}$

Now that the set of polynomials $F[X]_{p(X)}$ has been given the structure of a ring, it makes sense to ask whether this ring contains the reciprocal of each of its nonzero elements. A clever little algorithm can be used to show the existence of reciprocals, thereby giving an affirmative answer to our question.

The algorithm depends on the validity of the following lemma, which may seem paradoxical at first sight. It says that *you can reduce the degree of a polynomial by multiplying it by another polynomial!*

You must remember, however, that it is not the usual multiplication of polynomials which is involved, but rather multiplication "modulo $p(X)$".

8.2.1 Lemma. [Degree Reducing Lemma] *Assume $a(X)$ is a nonconstant polynomial in $\mathbb{F}[X]_{p(X)}$. If $p(X)$ is irreducible over $\mathbb{F}$ then there exists a polynomial $q(X)$ in $\mathbb{F}[X]_{p(X)}$ such that the polynomial*

$$a(X) \otimes_{p(X)} q(X)$$

is nonzero and has degree less than that of $a(X)$.

Proof. By the Division Theorem (Theorem 1.3.2), we can divide $p(X)$ by $a(X)$ to get a quotient $q(X)$ and remainder $r(X)$, both in $\mathbb{F}[X]$, such that

$$p(X) = a(X) q(X) + r(X) \qquad (*)$$

and either   (i) $r(X) = 0$
or   (ii) $\deg r(X) < \deg a(X)$.

But if (i) holds then, by $(*)$,

$$p(X) = a(X) q(X). \qquad (**)$$

Since, however, $a(X)$ is a nonconstant polynomial in $\mathbb{F}[X]_{p(X)}$,

$$0 < \deg a(X) < \deg p(X)$$

and so $(**)$ contradicts the irreducibility of $p(X)$. Thus (i) is false and hence

$$\boxed{r(X) \neq 0}$$

and (ii) holds. Thus

$$\boxed{\deg r(X) < \deg a(X).}$$

From $(*)$ it now follows that

$$\deg p(X) = \deg \big(a(X) q(X)\big)$$
$$= \deg a(X) + \deg q(X).$$

An Algebraic Postscript

This shows that $\deg q(X) < \deg p(X)$ and hence

$$\boxed{q(X) \in \mathsf{F}[X]_{p(X)}.}$$

Finally note that, by (∗), the remainder on dividing $a(X)\,q(X)$ by $p(X)$ is $-r(X)$. Hence

$$\boxed{a(X) \otimes_{p(X)} q(X) = -r(X).}$$

The conclusions of the lemma now follow from the boxed statements. ∎

Note that there is an algorithm implicit in the above proof: *To get the "degree reducing factor" $q(X)$ you divide the irreducible polynomial $p(X)$ by the polynomial $a(X)$ whose degree you wish to reduce. The quotient $q(X)$ and the remainder $r(X)$ then satisfy*

$$a(X) \otimes_{p(X)} q(X) = -r(X)$$

where $r(X)$ is nonzero and has lower degree than $a(X)$.

We are now able to give an algorithm for calculating reciprocals in $\mathsf{F}[X]_{p(X)}$. For brevity, we use the symbol $\otimes$ to stand for the multiplication $\otimes_{p(X)}$ in $\mathsf{F}[X]_{p(X)}$.

8.2.2 Algorithm. [Calculation of Reciprocals] Assume that $a(X)$ is a nonzero polynomial in the ring $\mathsf{F}[X]_{p(X)}$ where $p(X)$ is irreducible over F. Construct, using the Degree Reducing Lemma (Lemma 8.2.1), polynomials $q_1(X), q_2(X), q_3(X), \ldots$ in $\mathsf{F}[X]_{p(X)}$ such that

$$a(X) \otimes q_1(X),$$
$$a(X) \otimes q_1(X) \otimes q_2(X),$$
$$a(X) \otimes q_1(X) \otimes q_2(X) \otimes q_3(X),$$

and so on, have successively lower degrees. Hence eventually we get

$$a(X) \otimes q_1(X) \otimes q_2(X) \otimes q_3(X) \otimes \ldots \otimes q_m(X) = c,$$

where $c \neq 0$ is a constant in the field F. If we put

$$d(X) = \frac{1}{c} q_1(X) \otimes q_2(X) \otimes q_3(X) \otimes \ldots \otimes q_m(X),$$

then $d(X)$ is the reciprocal of $a(X)$ in $\mathsf{F}[X]_{p(X)}$.

Proof. It is easy to see that
$$a(X) \otimes d(X) = 1. \qquad \blacksquare$$

The polynomial $d(X)$ so constructed, being the reciprocal of $a(X)$ in the ring $\mathbb{F}[X]_{p(X)}$, is denoted by
$$1/a(X)$$
(although this notation does not indicate the dependence of the answer on the polynomial $p(X)$).

The following rough and ready description of the above algorithm for obtaining $1/a(X)$ may be helpful when you apply it to specific examples.

(A) *Find successive reducing factors for $a(X)$ until you get a constant.*

(B) *Multiply all the reducing factors together (modulo $p(X)$) and then divide by the constant. This gives you $1/a(X)$.*

8.2.3 Example. Find the reciprocal of $X^2 + 1$ in $\mathbb{Q}[X]_{X^3-2}$.

(A) We find successive reducing factors for $X^2 + 1$.

1. To find a reducing factor for $X^2 + 1$, divide this polynomial into $X^3 - 2$ to get
$$X^3 - 2 = (X^2 + 1)X + (-X - 2).$$
Hence
$$(X^2 + 1) \otimes_{X^3-2} X = X + 2.$$
Thus $X + 2$ is the result of applying the *reducing factor* X.

2. To find a reducing factor for $X + 2$, divide this polynomial into $X^3 - 2$ to get
$$X^3 - 2 = (X + 2)(X^2 - 2X + 4) - 10.$$
Hence
$$(X + 2) \otimes_{X^3-2} (X^2 - 2X + 4) = 10.$$
Thus the constant 10 is the result of applying a further *reducing factor* $X^2 - 2X + 4$.

An Algebraic Postscript

(B) *We multiply the reducing factors and divide by the constant.*
This gives

$$\frac{1}{X^2+1} = \frac{1}{10}X \otimes_{X^3-2} (X^2 - 2X + 4)$$
$$= -\frac{1}{5}X^2 + \frac{2}{5}X + \frac{1}{5}. \qquad \blacksquare$$

It now follows easily that $F[X]_{p(x)}$ is a field (provided $p(X)$ is irreducible over F).

8.2.4 Theorem. *If $p(X)$ is irreducible over F then $F[X]_{p(X)}$, with the operations in Definitions 8.1.1, is a field.*

Proof. We know from Section 8.1 that it is a ring. It is clearly commutative and contains 1. By Algorithm 8.2.2, it is closed under division by nonzero elements. Hence it is a field. $\blacksquare$

This gives us an important method for constructing new fields – find a polynomial $p(X)$ which is irreducible over a field F and construct $F[X]_{p(X)}$. For example, you may have thought that there could be no field with four elements since the obvious contender $\mathbb{Z}_4$ is not a field. We can, however, use the $F[X]_{p(X)}$ construction to exhibit a field with four elements.

8.2.5 Example. [A Field with Four Elements] We shall apply Theorem 8.2.4, with $F = \mathbb{Z}_2$ and $p(X) = X^2 + X + 1$.

Because $p(0) \neq 0$ and $p(1) \neq 0$, it follows from the Small Degree Irreducibility Theorem (Theorem 4.2.3) that $p(X)$ is irreducible over $\mathbb{Z}_2$. Now

$$E = \mathbb{Z}_2[X]_{p(X)} = \{a_0 + a_1 X : a_0, a_1 \in \mathbb{Z}_2\}$$

has just four elements $0, 1, X, 1+X$ and, by the above theorem, it is a field.

It is easy to draw up tables for the two operations in this field.

The tables are shown below.

$\oplus_{p(X)}$	0	1	X	1+X
0	0	1	X	1+X
1	1	0	1+X	X
X	X	1+X	0	1
1+X	1+X	X	1	0

$\otimes_{p(X)}$	0	1	X	1+X
0	0	0	0	0
1	0	1	X	1+X
X	0	X	1+X	1
1+X	0	1+X	1	X

Note also that (as you might expect from Theorem 3.2.4)
$$[\mathsf{E} : \mathbb{Z}_2] = \deg p(X),$$
each of these numbers being equal to 2. ∎

─────────── **Exercises 8.2** ───────────

1. In the ring $\mathbb{Q}[X]_{X^3-2}$ find successive reducing factors for the polynomial X^2+X+1 and hence find the reciprocal of this element.

2. (a) In the ring $\mathbb{R}[X]_{X^2+1}$ find
    (i) $1/(3+4X)$,
    (ii) $1/(a+bX)$, if a and b are not both zero.
   (b) Does your answer to (a)(ii) agree with the formula for $1/(a+bi)$ in $\mathbb{C}$? (See part (c) of Exercises 8.1 #2.)

3. Which of the following are fields? (Justify your answers.)
   (a) $\mathbb{Q}[X]_{X^3-2X+1}$,
   (b) $\mathbb{Q}[X]_{X^3+2X+1}$,
   (c) $\mathbb{Q}(\sqrt[3]{2})[X]_{X^3-2}$.

An Algebraic Postscript

4. In Example 8.2.3, check the very last step of the calculation and then verify directly, using multiplication modulo $X^3 - 2$, that the answer given for the reciprocal of $1 + X^2$ is indeed correct.

5. Find a polynomial of degree 2 in $\mathbb{Z}_3[X]$ which is irreducible over $\mathbb{Z}_3$. Follow the method of Example 8.2.5 to construct a field with nine elements.

6. Show how to construct a field with eight elements.

   [Hint. You will need a polynomial of degree three in $\mathbb{Z}_2[X]$ which is irreducible over $\mathbb{Z}_2$.]

8.3 Reciprocals in $\mathbb{F}(\alpha)$

We return to the situation in the earlier chapters where α is a complex number which is algebraic over a subfield $\mathbb{F}$ of $\mathbb{C}$, with

$$\text{irr}(\alpha, \mathbb{F}) = p(X) \quad \text{and} \quad \deg(\alpha, \mathbb{F}) = n.$$

Although we were able to prove in Chapter 3 that $\mathbb{F}(\alpha)$ is a field, we had no efficient way of expressing $1/\beta$ in the form

$$c_0 + c_1\alpha + \ldots + c_{n-1}\alpha^{n-1} \quad (c_i \in \mathbb{F})$$

for an arbitrary nonzero β in $\mathbb{F}(\alpha)$. We can now do this very easily, however, by using the algorithm from Section 8.2.

If you look at the definition of $\mathbb{F}(\alpha)$ (Definition 3.2.1) and at the definition of $\mathbb{F}[X]_{p(X)}$ in Section 8.1, you will notice a certain similarity. We bring this out by defining the *evaluation mapping*

$$\phi_\alpha : \mathbb{F}[X]_{p(X)} \longrightarrow \mathbb{F}(\alpha)$$

by putting

$$\phi_\alpha(c_0 + c_1 X + \ldots c_{n-1}X^{n-1}) = c_0 + c_1\alpha + \ldots + c_{n-1}\alpha^{n-1}.$$

Note that ϕ_α simply replaces the indeterminate X by the number α.

Recall that a field isomorphism is a one-to-one and onto mapping between two fields which preserves the addition and multiplication operations. Such an isomorphism also preserves subtraction and division. Two isomorphic fields are essentially the same, except for the names of their elements.

8.3.1 Proposition. *The map ϕ_α is a field isomorphism.*

Proof. That ϕ_α is a one-to-one and onto map follows routinely from the fact that $\{1, \alpha, \ldots, \alpha^{n-1}\}$ is a basis for $\mathbb{F}(\alpha)$ over $\mathbb{F}$ (see Exercises 8.3 #8). It is also easy to show that ϕ_α preserves addition in the sense that

$$\phi_\alpha\big(f(X) \oplus_{p(X)} g(X)\big) = \phi_\alpha\big(f(X)\big) + \phi_\alpha\big(g(X)\big)$$

for all polynomials $f(X)$ and $g(X)$ in $\mathbb{F}[X]$.

What about multiplication? This is a bit more complicated. By the Division Theorem (Theorem 1.3.2), there exist polynomials $q(X)$ and $r(X)$ in $\mathbb{F}[X]$ with

$$f(X)\, g(X) = p(X)\, q(X) + r(X) \tag{1}$$

where $r(X) = 0$ or $\deg r(X) < \deg p(X)$. If we replace X by α in (1) we get

$$\begin{aligned} f(\alpha)\, g(\alpha) &= p(\alpha)\, q(\alpha) + r(\alpha) \\ &= 0 \cdot q(\alpha) + r(\alpha) \\ &= r(\alpha). \end{aligned} \tag{2}$$

But by the definition of multiplication in the field $\mathbb{F}[X]_{p(X)}$,

$$r(X) = f(X) \otimes_{p(X)} g(X).$$

Hence in the equation (2),

$$\text{left hand side} = \phi_\alpha\big(f(X)\big) \cdot \phi_\alpha\big(g(X)\big)$$

while

$$\text{right hand side} = \phi_\alpha\big(f(X) \otimes_{p(X)} g(X)\big).$$

These two sides are equal, which implies that the map ϕ_α preserves multiplication. ∎

We can now put the algorithm for reciprocals in $\mathbb{F}[X]_{p(X)}$ to work in $\mathbb{F}(\alpha)$ since these two fields are isomorphic.

8.3.2 Example. Find the reciprocal of the number $1 + (\sqrt[3]{2})^2$ in the field $\mathbb{Q}(\sqrt[3]{2})$.

An Algebraic Postscript

Note firstly that, because

$$\mathrm{irr}(\sqrt[3]{2}, \mathbb{Q}) = X^3 - 2,$$

it follows that $\phi_{\sqrt[3]{2}}$ is an isomorphism from $\mathbb{Q}[X]_{X^3-2}$ to $\mathbb{Q}(\sqrt[3]{2})$. Under this isomorphism,

$$1 + (\sqrt[3]{2})^2 \in \mathbb{Q}(\sqrt[3]{2}) \text{ corresponds to } 1 + X^2 \in \mathbb{Q}[X]_{X^3-2}$$

or, more formally,

$$1 + (\sqrt[3]{2})^2 = \phi_{\sqrt[3]{2}}(1 + X^2).$$

Hence

$$\frac{1}{1 + (\sqrt[3]{2})^2} = \frac{1}{\phi_{\sqrt[3]{2}}(1 + X^2)}$$

$$= \phi_{\sqrt[3]{2}}\left(\frac{1}{1 + X^2}\right) \quad \text{as } \phi_{\sqrt[3]{2}} \text{ is an isomorphism}$$

$$= \phi_{\sqrt[3]{2}}\left(-\frac{1}{5}X^2 + \frac{2}{5}X + \frac{1}{5}\right) \quad \text{by Example 8.2.3}$$

$$= -\frac{1}{5}(\sqrt[3]{2})^2 + \frac{2}{5}(\sqrt[3]{2}) + \frac{1}{5}. \qquad \blacksquare$$

This concludes our algebraic postscript in which we have shown how to construct new fields (including finite ones) and have given an efficient way of calculating reciprocals in fields of the form $\mathbb{F}[X]_{p(X)}$ and $\mathbb{F}(\alpha)$.

———————————— **Exercises 8.3** ————————————

1. For which choice of the polynomial $p(X)$ is the field $\mathbb{Q}(\sqrt[3]{2})$ isomorphic to the field $\mathbb{Q}[X]_{p(X)}$?

2. Check (using multiplication in $\mathbb{Q}(\sqrt[3]{2})$) the answer obtained in Example 8.3.2.

3. (a) Write out explicitly each of the sets $\mathbb{Q}(\sqrt[3]{2})$ and $\mathbb{Q}[X]_{X^3-2}$.
   (b) Write down (i) a typical element of the latter set,
   (ii) its image under the map $\phi_{\sqrt[3]{2}}$,
   and show both in the appropriate sets in the following diagram.

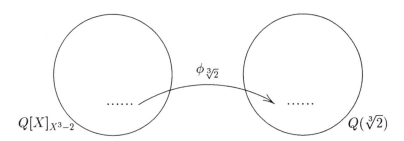

(c) Which theorem from Chapter 3 guarantees that each element of $\mathbb{Q}(\sqrt[3]{2})$ is *uniquely expressible* in the form
$$c_0 + c_1\sqrt[3]{2} + c_2(\sqrt[3]{2})^2$$
with each c_i in $\mathbb{Q}$?

(d) Apply the map $\phi_{\sqrt[3]{2}}^{-1}$ (the inverse of the map $\phi_{\sqrt[3]{2}}$) to the above element.

4. Let $f(X) = X^2 + 1$ and $g(X) = 2X + 3$, which are elements of $\mathbb{Q}[X]$.
   (a) State why $f(X)$ and $g(X)$ belong to $\mathbb{Q}[X]_{X^3-2}$.
   (b) Calculate the following sum and product in the ring $\mathbb{Q}[X]_{X^3-2}$.
   (i) $f(X) \oplus_{X^3-2} g(X)$,
   (ii) $f(X) \otimes_{X^3-2} g(X)$.
   (c) Use the fact that $\phi_{\sqrt[3]{2}}$ is a homomorphism from $\mathbb{Q}[X]_{X^3-2}$ to $\mathbb{Q}(\sqrt[3]{2})$ to calculate its value at the product in (b)(ii). Check your answer by applying the map directly to your answer for (b)(ii).

5. Use your answer to Exercises 8.2 #1 to find the reciprocal in $\mathbb{Q}(\sqrt[3]{2})$ of the number $(\sqrt[3]{2})^2 + \sqrt[3]{2} + 1$. Indicate where you use the fact that $\phi_{\sqrt[3]{2}}$ is an isomorphism.

6. Repeat Exercise 5, but this time follow the method used in the proof of Theorem 3.2.6 and in Example 3.2.7. Which method seems the more efficient, this one or that of Exercise 5?

7. Use your answer to Exercise 5 to express
$$\left(1 + 6\sqrt[3]{2} - 2(\sqrt[3]{2})^2\right) / \left(1 + \sqrt[3]{2} + (\sqrt[3]{2})^2\right)$$
in the form $c_0 + c_1\sqrt[3]{2} + c_2(\sqrt[3]{2})^2$ for $c_0, c_1, c_2 \in \mathbb{Q}$.

8. Prove that the map ϕ_α defined in the text is one-to-one and onto. [Hint. Use the fact that $\{1, \alpha, \ldots, \alpha^{n-1}\}$ is a basis for $\mathsf{F}(\alpha)$ over F.]

Additional Reading for Chapter 8

The field $\mathsf{F}[X]_{p(X)}$ is usually constructed as a factor ring of the ring $\mathsf{F}[X]$ of polynomials. See, for example, Section 35 of [JF] and Chapter 10 of [WG].

Reciprocals of elements in $\mathsf{F}(\alpha)$ are often calculated using the algorithm for finding the greatest common divisor of two polynomials, as indicated, for example, in Section 110 of [AC] and Section 2.2 of [CH].

Finite fields have been completely characterized: see, for example, Section 45 of [JF] and Chapter 11 of [WG].

[AC] A. Clark, *Elements of Abstract Algebra*, Wadsworth, Belmont, California, 1971.

[JF] J.B. Fraleigh, *A First Course in Abstract Algebra*, 3rd edition, Addison-Wesley, Reading, Massachusetts, 1982.

[WG] W.J. Gilbert, *Modern Algebra with Applications*, Wiley, New York, 1976.

[CH] C.R. Hadlock, *Field Theory and its Classical Problems*, Carus Mathematical Monographs, No. 19, Mathematical Association of America, 1978.

CHAPTER 9

Other Impossibilities and Abstract Algebra

This book has introduced some basic ideas from the part of abstract algebra which deals with fields. Use of these ideas made it relatively easy for us to prove the impossibility of the constructions mentioned in the Introduction.

It may now come to you as a pleasant surprise that your newly acquired ideas about fields have further applications in proving the impossibility of solving other famous problems. Three such problems are discussed below. While the first problem comes from geometry, the other two come from elementary algebra and elementary calculus respectively.

9.1 Construction of Regular Polygons

A polygon is said to be *regular* if its sides all have the same length and its angles are all equal. Constructing a regular polygon with n sides amounts to dividing an angle of 360° into n equal parts.

The ancient Greeks were able to construct, with straightedge and compass, regular polygons with 3 sides and 5 sides, but were not able to construct one with 7 sides. It was, of course, trivial for them to construct a polygon with 2^m sides, for any integer $m \geq 2$, as this involves merely repeated bisection. They also proved that if it is possible to construct regular polygons with r sides and s sides, then it is also possible to construct one with rs sides, provided that the integers r and s have greatest common divisor 1. Thus, taking $r = 3$ and $s = 5$, they knew it is possible to construct one with 15 sides.

No further progress was made on the problem of constructing regular polygons for over 2000 years until in 1796 the nineteen year old *Carl Friederich Gauss* amazed the mathematical world by constructing a regular polygon with 17 sides. He did this by showing that the equation $x^{17} - 1 = 0$ can be reduced to a finite set of quadratic equations. So the regular polygon with 17 sides can be constructed with straightedge and compass. Gauss showed, furthermore, that a regular polygon with n sides can be constructed whenever n is a prime number of the form

$$n = 2^{2^k} + 1$$

for some integer $k \geq 0$. Prime numbers of this form are called *Fermat primes*. (Notice that $17 = 2^{2^2} + 1$ is a Fermat prime.)

From the results mentioned above, it follows that a regular polygon with n sides can be constructed if

$$n = 2^{i+2} \quad \text{or} \quad n = 2^i p_1 p_2 \ldots p_j$$

where $i \geq 0$ is an integer and $p_1, p_2, \ldots, p_j$ are distinct Fermat primes. What if n does not have this form? The answer was provided by *Pierre Wantzel* in 1837, who proved that, for n not of this form, the construction of a regular polygon with n sides in impossible. The proof of this impossibility requires little more than the ideas about fields already given in this book. Books which cover this topic in detail are listed in the "Additional Reading" at the end of this chapter.

9.2 Solution of Quintic Equations

Our next problem is concerned with solving polynomial equations, that is, equations of the form

$$a_n x^n + a_{n-1} x^{n-1} + \ldots + a_1 x + a_0 = 0 \qquad (a_i \in \mathbb{R})$$

such as, for example,

$$2x^3 - 5x^2 + 2x - 11 = 0.$$

Such equations arose early in the history of mathematics: the ancient Babylonians solved many particular quadratic equations, while the Arabs discovered a general formula for the solutions during the Middle Ages. Their formula is equivalent to the one used today by school children to solve $ax^2 + bx + c = 0$:

$$x = \frac{-b \pm \sqrt{b^2 - 4ac}}{2a} \qquad (a \neq 0).$$

During the Renaissance, Italian mathematicians discovered formulae for solving cubic and quartic equations. Thus, for example, to solve a cubic of the special form $x^3 + px = q$, they used a formula which, in modern notation, reads

$$x = \sqrt[3]{\frac{q}{2} + \sqrt{\frac{q^2}{4} + \frac{p^3}{27}}} + \sqrt[3]{\frac{q}{2} - \sqrt{\frac{q^2}{4} + \frac{p^3}{27}}}.$$

Cases in which the quantity under the square root sign is negative puzzled them, however, as complex numbers were not properly understood at that time.

The above formulae are said to provide *solutions by radicals* of the original equations. The idea is that the solutions are obtained from the coefficients in the equations by applying successive field operations and extractions of roots.

Attempts were made to find similar formulae to solve an arbitrary quintic equation, but none of these attempts succeeded. In 1824, *Niels Henrick Abel*, after mistakenly believing that he had found the solution, proved that it is **impossible** to solve an arbitrary quintic by radicals. Soon afterwards, the youthful genius *Évariste Galois* established a necessary and sufficient condition for a polynomial of arbitrary degree to be solvable by radicals. He was then able to extend

Abel's result and to show that, for each integer $n \geq 5$, it is **impossible** to solve an arbitrary equation of degree n by radicals.

Although a complete discussion of these results would require a lot more abstract algebra than has been developed in this book, it is possible to indicate how the problem can be expressed in terms of field extensions.

To this end, let $f(X)$ be a polynomial in $\mathbb{F}[X]$ where $\mathbb{F}$ is a subfield of $\mathbb{C}$. Instead of referring to the equation $f(x) = 0$ as being "solvable by radicals", we apply this phrase to the polynomial $f(X)$ itself, the precise definition being as follows:

9.2.1   Definition. *A polynomial $f(X) \in \mathbb{F}[X]$ is said to be solvable by radicals over $\mathbb{F}$ if there is a sequence of complex numbers $c_1, c_2, c_3, \ldots, c_n$ and a sequence of positive integers $i_1, i_2, i_3, \ldots, i_n$ such that*

$$c_1^{i_1} \in \mathbb{F}$$
$$c_2^{i_2} \in \mathbb{F}(c_1)$$
$$c_3^{i_3} \in \mathbb{F}(c_1)(c_2)$$
$$\vdots$$
$$c_n^{i_n} \in \mathbb{F}(c_1)(c_2) \ldots (c_{n-1})$$

and the zeros in $\mathbb{C}$ of $f(X)$ are all contained in the field

$$\mathbb{F}(c_1)(c_2) \ldots (c_{n-1})(c_n). \qquad \blacksquare$$

This definition has been expressed in terms of integer powers of complex numbers in order to avoid the ambiguity which would arise from the use of roots of complex numbers.

The following theorem expresses, in a precise way, the fact that it is impossible to solve an arbitrary quintic or higher degree equation by radicals.

9.2.2   Theorem. *There exist polynomials of every degree $n \geq 5$ which are not solvable by radicals over $\mathbb{Q}$.*

The proof consists of choosing suitable polynomials and verifying that they are not solvable by radicals. The choice given in [CH] is $2X^5 - 5X^4 + 5$ for $n = 5$ and, for $n \geq 5$, the result of multiplying this polynomial by X^{n-5}. The proof that these polynomials are not solvable by radicals, however, involves use of the field and group theory developed by Galois. Books which cover this topic in detail are listed in the "Additional Reading" at the end of this chapter.

9.3 Integration in Closed Form

A topic which in the past has tended to dominate the undergraduate curriculum in elementary calculus is that of finding indefinite integrals in "closed form"; that is, in terms of functions which are usually defined in introductory calculus courses.

This topic is usually presented as an art rather than as a science: each integral comes out if the student is clever enough (or lucky enough!) to guess a suitable trick. There are, however, some integrals which look quite innocent and yet refuse to "come out". One such example is

$$\int e^{x^2} dx.$$

The impossibility of finding this integral in closed form was proved by *Joseph Liouville*, who, in the years 1833–1841, laid the foundations for the study of integration as a science, rather than an art.

At first sight one might guess that, of all the topics in the undergraduate curriculum, none could be further removed from abstract algebra than that of finding integrals in closed form – not so, however! In the "suggested reading" section of [MS], written in the mid 1960's, Spivak mentions that polished algebraic treatments of Liouville's theorems had recently been obtained. One such treatment is given in [MR], which includes, in particular, an algebraic proof of the impossibility of finding the above integral in closed form. Even the most cursory reading of [MR] will reveal its dependence on ideas about fields similar to those developed earlier in this book:

algebraic and transcendental elements,

extensions of fields,

degree of one field over another,

irreducible polynomials of elements,

towers of fields.

A full understanding of [MR] requires a knowledge of functions of a complex variable, because the elements of the fields involved are no longer numbers, but functions defined on subsets of $\mathbb{C}$. As one might guess, fields are involved at the very outset, in setting up the problem in a precise way.

The most concrete evidence of the recent work in this area can be seen in a number of computer algebra packages including Reduce™, Mathematica™ and MACSYMA™ which are available on various computers ranging from microcomputers to mainframes. These packages contain implementations of the algorithm due to R.H. Risch [RR] for integrating an important class of functions in closed form. When you type in one of these functions the package tells you whether or not the function can be integrated in closed form and, if so, displays the answer.

Additional Reading for Chapter 9

The constructibility of regular polygons is studied in Sections 135–138 of [AC], Section 48 of [JF] and Section 10.2 of [LS]. Each of these books gives an account of Galois theory, while Sections 139–149 of [AC] give an extensive account of the application of Galois theory to the solution by radicals of polynomial equations of low degree. A recent re-examination of the short, but action-packed, life of Galois can be found in [TR].

The article by A.C. Norman in [CA] gives an account of recent algorithms for integration in closed form, suitable for implementation on a computer, while [JD] presents the underlying mathematical theory.

[CA] *Computer Algebra: Symbolic Algebraic Computation,* ed. B. Buchberger, G.E. Collins, R. Loos and R. Albrecht, Springer-Verlag, Vienna, 2nd edition, 1983.

[AC] A. Clark, *Elements of Abstract Algebra,* Wadsworth, Belmont, California, 1971.

[JD] J.H. Davenport, *On the Integration of Algebraic Functions,* Lecture Notes in Computer Science, 102, Springer-Verlag, Berlin, 1981.

[JF] J.B. Fraleigh, *A First Course in Abstract Algebra,* 3rd edition, Addison-Wesley, Reading, Massachusetts, 1982.

[CH] C.R. Hadlock, *Field Theory and its Classical Problems,* Carus Mathematical Monographs, No. 19, Mathematical Association of America, 1978.

[MR] M. Rosenlicht, "Integration in Finite Terms", *American Mathematical Monthly,* 79 (1972), 963–972.

[TR] T. Rothman, "Genius and Biographers: the Fictionalization of Évariste Galois", *American Mathematical Monthly,* 89 (1982), 84–106.

[RR] R.H. Risch, "The Solution of the Problem of Integration in Finite Terms", *Transactions of the American Mathematical Society,* 76 (1970), 605–608.

[LS] L.W. Shapiro, *Introduction to Abstract Algebra,* McGraw-Hill, New York, 1975.

[MS] M. Spivak, *Calculus,* W.A. Benjamin, New York, 1967.

Index

abelian group, 24–25

abstract algebra, 3

Ahmes, 2

algebra, i, 3, 7, 75, 115

algebraic, i–ii, 7, 14, 75, 99, 103, 105, 181

algebraic number, 27–28, 30–31, 33, 72, 99, 147, 153, 157

algebraic over a field, 27–33, 39, 42, 44, 48, 50–51, 68, 72, 101–102, 124, 147, 171

Algorithm

—Calculation of Reciprocals, 167

—Division, 17–19

—Euclidean, i

—Fundamental Algorithm for Symmetric Functions, 139

All Constructibles Come from Square Roots Theorem, 96, 100–101, 108, 110

approximation, 121, 123, 125

Archimedes, 2

basis, i, 11, 26, 40, 52–53, 55–56

Basis for $F(\alpha)$ Theorem, 46

Basis for a Tower Theorem, 53

Binomial Theorem, 107, 119

bisect, 3, 77–78, 80, 91

bisecting a line segment. *See* Standard Constructions.

bisecting an angle. *See* Standard Constructions.

C, 8

Calculation of Reciprocals, 167

calculus, 115, 117, 149, 177, 181

coefficient, 13–15, 137, 144. *See also* leading coefficient, equating coefficients.

—integer, 120, 128, 130, 147

—rational, 148

commutative, 10

commutative group, 24

commutative ring, 13, 24–25

compass, 1, 3–4, 75–76, 78, 81. *See also* straightedge and compass.

—collapsing, 82

—noncollapsing, 82

—rusty, 85

complex analysis, 150

complex number, 8, 16–17, 146, 156

complex-valued function, 115, 149–150

complex-valued integral, 149

computer algebra, 182

CON, 93–94

constant, 69

constant polynomial, 63–65, 70

constructible number, 4, 75, 93–96, 99, 101, 103–104, 110

constructing a line parallel to a given line. *See* Standard Constructions.

constructing a product. *See* Standard Constructions.

constructing a quotient. *See* Standard Constructions.

constructing a square root. *See* Standard Constructions.

constructing an angle of 60° or 90°. *See* Standard Constructions.

Construction Rules, 89

—rule (i), 90

—rule (ii), 90

copying an angle. *See* Standard Constructions.

cubic, 4, 68, 104

De Moivre's Theorem, 107

deg $f(X)$, 14

deg(α, F), 33

degree, 138

—least, 32–34, 62, 70

—of a number over a field, 33, 42, 46, 99, 101

—of divisor, 18

—of extension field, 12, 181

—of polynomial, 14, 17, 20, 46, 61–62, 118, 120, 138, 143, 145

—of remainder, 18

Degree of a Constructible Number Theorem, 101

Degree Reducing Lemma, 166–167

Delian problem, 1

dense, 121

dependent. *See* linearly dependent.

derivative, 149–150

differentiation, 118, 121

dimension, i, 4, 11–12, 26, 40, 46, 52, 56, 165

Dimension for a Tower Theorem, 55, 102

dividend, 18–19

division, 10, 165, 169, 171

—of polynomials, 17

Division Algorithm, 17–19

Division Theorem, 19, 65, 166, 172

divisor, 18–19

doubling the cube, 1, 61, 91, 93, 99, 104

duplicating the cube, 1. *See also* doubling the cube.

e, ii, 28, 115, 122–125, 131–132, 160

e^z, 146

Eisenstein's Irreducibility Criterion, 107

elementary symmetric function, 136–138

equating coefficients, 121, 128

Eratosthenes, 1

Euclid, 117

Euclidean Algorithm, i

expansion in powers, 119–120, 127–129

exploration, ii

exponential function, 118, 122, 126

extension field, 8, 71, 74, 143, 181

extension of F by α, 44

$\mathsf{F}[X]_{p(X)}$, 164

F-circle, 108–109, 111

F-line, 108–109, 111

F-point, 108–109, 111

factor, 20–21, 23, 56, 62, 116–117, 129, 131, 159

—common, 20–21, 23

—of degree 1, 64–65

Factor Theorem, 65, 67

factorize, 4, 61–63

Fermat prime, 178

field, i, 8–11, 24–25, 39–44, 46–47, 50, 56, 93–94, 163–165, 169, 177

—finite, 9, 13

—smallest, 41, 48

field of scalars, 10

Field with Square Roots Theorem, 93, 96

finite field, 9, 13

finite-dimensional, 4, 11–12, 55, 71–72

finite-dimensional extension, 71–72

Fundamental Algorithm for Symmetric Functions, 139–142, 145

Index 185

Fundamental Theorem of Algebra, 146–147
Fundamental Theorem of Arithmetic, 21, 23, 116–117
Fundamental Theorem of Calculus, 151
Fundamental Theorem on Symmetric Functions, 138–139, 143, 158, 160

geometrical, 99
geometrical constructions, 16, 75
—collapsing compass, 84
—doubling the cube, 2, 91, 93, 99, 104
—quintisecting an angle, 105
—regular polygon, 2
—rules for, 75, 89, 91
—squaring the circle, 2, 92, 105
—straightedge and compass, 3, 76, 81–83, 104–105
—three famous, 66, 75, 103
—trisecting an angle, 3, 83, 92, 104
geometry, i, 3, 75
—Euclidean, 77
Greek, 1–2, 4, 75, 81–82, 89
group, 24

higher order, 139

identity, 24–25
—additive, 24–25
—multiplicative, 25
impossibility, i–ii, 3–4, 61, 66, 74–75, 95–96, 99, 101, 105, 112, 115, 159, 163, 177–179, 181
independent. *See* linearly independent.
indeterminate, 14, 134, 138
induction. *See* mathematical induction.
initial set of points, 89–90, 93
integral, 150, 153
integral domain, 13, 25
integration, 115, 126, 181

—closed form, 181
—linearity of, 117, 126–127, 151
integration by parts, 118, 126, 151
inverse, 24
$\text{irr}(\alpha, \mathbf{F})$, 33
irrational number, 7, 20, 23, 122–123, 125, 133
irreducible, 34, 61, 63, 68–69, 164–166, 169
irreducible polynomial, 62–63, 68, 125
—over subfield, 63
irreducible polynomial of a number, 33–34, 42–43, 61–62, 68, 104, 125, 147, 181
isomorphism, 171, 173

leading coefficient, 14, 31, 137. *See also* coefficient.
limit, 119, 130, 156
Lindemann, 2, 160–161
linear algebra, i
linear combination, 11, 119
—trivial, 11, 26
linear span. *See* span.
linearity of integration. *See* integration.
linearly dependent, 11, 26, 30, 36, 47, 72
linearly independent, i, 11, 26, 36, 40, 46, 54–55
lower order, 139, 141
lowest terms, 20

MACSYMA, 182
Mathematica, 182
mathematical induction, 111, 126, 133, 141, 143, 154, 160
midpoint, 77
modulo, 164, 166
modulus, 152
Monic Irreducible Zero Theorem (MIZT), 68–69

monic polynomial, 31–34, 68–70, 143, 147–148
monomial, 18, 138, 141

N, 10
non-constructible numbers, 100

one-to-one mapping, 134
onto mapping, 134

π, i–ii, 2–5, 28, 92, 105, 115–116, 134, 143, 146–147, 149, 153, 156–157, 160–161
pairs of fields, 11
permutation, 134, 136
—identity, 135
polynomial, 7, 13–15, 118. *See also* monic polynomial.
—of degree 2 or 3, 65
—over a ring, 13–14
polynomial form, 13–15, 118
polynomial function, 13, 15, 118, 126
polynomial ring, 14–15
prime, 116–117, 127–129, 157, 165, 178
Problem I, 1–4, 104. *See also* doubling the cube.
Problem II, 1–4, 104. *See also* trisecting an angle.
Problem III, 1–4, 99, 115. *See also* squaring the circle.
proofs, i–ii, 4
Ptolemy, 1

Q, 8
quadratic, 61, 110
quadrature, 1–2
quintic equation, 179
quintisecting an angle, 105
quotient, 17–19, 166–167

R, 8

$R[X]$, 14
rational number, 4, 8, 20–21, 93, 121, 123, 125
Rational Roots Test (RRT), 7, 20–21, 67–68
real number, 4, 8, 21, 94, 121, 123
real-valued function, 149, 151
reciprocal, 40, 44, 163, 165, 167, 170–172
Reduce, 182
reduced, 34, 61
reducible, 62–63, 165
reducing factor, 167–170
regular polygon, 1–2, 178, 182
remainder, 17–19, 164, 166–167
Rhind Papyrus, 2
ring, 8, 10, 13–16, 24, 40, 44, 46–47, 164, 169
—finite, 17
ring of integers, 25
ring with unity, 24–25
ruler, 1, 3

scalar, 10–11, 29, 40, 46–47, 52–53
scalar multiplication, 10–12, 25
schizophrenic, 53
Small Degree Irreducibility Theorem (SDIT), 66–69
Smallest Field Theorem, 48
solution by radicals, 179, 182
solvable by radicals, 179–180
span, 25, 40, 44, 46, 50–51, 53–55
square roots, 100, 111
squaring the circle, i, 1, 4, 92, 99, 105, 115, 159, 161
Standard Constructions, 76
—angle of 60°, 79, 104
—angle of 90°, 79
—bisecting a line segment, 77
—bisecting an angle, 78

—copying an angle, 80
—line parallel to a given line, 81
—operations of type (i), 90–93, 109
—operations of type (ii), 90–93, 109
—product, 85
—quotient, 86
—square root, 87
—transferring a length, 78
straightedge, 3–4, 75–77, 80–81
straightedge and compass, 104–105, 159
subfield, 8–9, 11, 16, 25, 39–40, 44, 52, 93
—smallest, 8, 44
subring, 25, 40, 44, 47
subspace, 25, 40, 44–45
Successive Square Roots Give Constructibles Theorem, 95–96
symmetric polynomial, 115, 134–136, 138, 141, 143, 146–148, 156, 158. *See also* elementary symmetric function.
symmetry, 134

Taylor series, 122
term, 139
term of highest order, 139, 141
tetragonidzein, 2
three-dimensional, 11

tower of fields, 51–52, 55, 95, 102, 181
transcendental, i–ii, 4, 28, 115, 124, 132, 153, 181
transcendental number, i, 30, 124, 131, 157
transferring a length. *See* Standard Constructions.
trisecting an angle, 1–4, 92, 104
trivial linear combination, 11, 26

uniqueness, 127, 129, 154

vector, 10, 25, 29, 40, 46, 51–53
vector addition, 10–11, 25
vector space, i, 4, 8, 10–11, 24–26, 29, 39–40, 46–47, 51–56, 165
vector subspace. *See* subspace.

Wantzel, 2

$\mathbf{Z}$, 10
$\mathbf{Z}_n$, 9
zero of a polynomial, 20–21, 28–29, 31, 34, 64–66, 68–69, 137, 143, 145, 147–148
zero polynomial, 14, 118
zero vector, 11, 25–26
zero-divisor, 10, 13, 24–25

Universitext   *(continued)*

Samelson: Notes on Lie Algebras
Smith: Power Series From a Computational Point of View
Smoryński: Logical Number Theory I: An Introduction
Smoryński: Self-Reference and Modal Logic
Stroock: An Introduction to the Theory of Large Deviations
Sunder: An Invitation to von Neumann Algebras
Tondeur: Foliations on Riemannian Manifolds
Verhulst: Nonlinear Differential Equations and Dynamical Systems
Zaanen: Continuity, Integration and Fourier Theory